T0296110

LONDON MATHEMATICAL SOCIETY LECTURE NOTE SERIES

Managing Editor: Professor M. Reid, Mathematics Institute,
University of Warwick, Coventry CV4 7AL, United Kingdom

The titles below are available from booksellers, or from Cambridge University Press at
www.cambridge.org/mathematics

London Mathematical Society Lecture Note Series: 430

Recent Progress in the Theory of the Euler and Navier–Stokes Equations

Edited by

JAMES C. ROBINSON
University of Warwick

JOSÉ L. RODRIGO
University of Warwick

WITOLD SADOWSKI
Uniwersytet Warszawski, Poland

ALEJANDRO VIDAL-LÓPEZ
Xi'an Jiaotong-Liverpool University, P. R. China

CAMBRIDGE
UNIVERSITY PRESS

CAMBRIDGE
UNIVERSITY PRESS

University Printing House, Cambridge CB2 8BS, United Kingdom

Cambridge University Press is part of the University of Cambridge.

It furthers the University's mission by disseminating knowledge in the pursuit of education, learning and research at the highest international levels of excellence.

www.cambridge.org
Information on this title: www.cambridge.org/9781107554979

© Cambridge University Press 2016

This publication is in copyright. Subject to statutory exception and to the provisions of relevant collective licensing agreements, no reproduction of any part may take place without the written permission of Cambridge University Press.

First published 2016

A catalogue record for this publication is available from the British Library

ISBN 978-1-107-55497-9 Paperback

Cambridge University Press has no responsibility for the persistence or accuracy of URLs for external or third-party internet websites referred to in this publication, and does not guarantee that any content on such websites is, or will remain, accurate or appropriate.

To our families

Contents

Preface

This volume is the result of a workshop, "The Navier-Stokes Equations in Venice", which took place in the Palazzo Pesaro-Papafava in Venice (part of the *Warwick in Venice* program), April 8th–12th, 2013.

Several of the speakers agreed to write review papers related to their contributions to the workshop, while others have written more traditional research papers. We believe that this volume therefore provides an accessible summary of a wide range of active research topics, along with some exciting new results, and we hope that it will prove a useful resource for both graduate students new to the area and to more established researchers.

We would like to express their gratitude to the following sponsors of the workshop and the writing of this volume of proceedings: JCR was supported by an EPSRC Leadership Fellowship (grant EP/G007470/1). JLR is currently supported by the European Research Council (ERC grant agreement n. 616797).

Finally it is a pleasure to thank Chiara Croff, the Venice administrator of *Warwick in Venice*, for her assistance during the organization of the workshop.

James C. Robinson
José L. Rodrigo
Witold Sadowski
Alejando Vidal López

Contributors

Those contributors who presented their work at the Warwick in Venice meeting are indicated by a star in the following list. We have also included the addresses of the editors.

Hugo Beirão da Veiga*
Dipartimento di Matematica,
Università di Pisa,
Via F. Buonarroti 1/c, Pisa. Italy.
bveiga@dma.unipi.it

Zachary Bradshaw
Department of Mathematics,
University of British Columbia,
1984 Mathematics Road,
Vancouver, B.C.
Canada V6T 1Z2
zbradshaw@math.ubc.ca

Weiwei Hu
Department of Mathematics,
University of Southern California,
Los Angeles, CA 90089. USA.
weiweihu@usc.edu

Giovanni P. Galdi*
Department of Mechanical Engineering and Materials Science,
University of Pittsburgh,
Pittsburgh, PA 15261. USA.
galdi@pitt.edu

John D. Gibbon*
Department of Mathematics,
Imperial College London,
London, SW7 2AZ. UK.
j.d.gibbon@ic.ac.uk

Zoran Grujić
Department of Mathematics,
Kerchof Hall,
University of Virginia,
Charlottesville, VA 22904. USA.
zg7c@virginia.edu

Igor Kukavica*
Department of Mathematics,
University of Southern California,
Los Angeles, CA 90089. USA.
kukavica@usc.edu

Adam Larios
Department of Mathematics,
University of Nebraska-Lincoln,
Lincoln, NE 68588–0130. USA.
alarios@unl.edu

Pierre Gilles Lemarié–Rieusset*
Laboratoire de Mathématiques et Modélisation d'Évry,
UMR CNRS 8071,
Université d'Évry.
France.
plemarie@univ-evry.fr

Giusy Mazzone
Department of Mechanical Engineering and Materials Science,
University of Pittsburgh,
Pittsburgh, PA 15261. USA.
gim20@pitt.edu

Benjamin Pooley
Mathematics Institute, University of Warwick,
Coventry, CV4 7AL. UK.
B.C.Pooley@warwick.ac.uk

James C. Robinson
Mathematics Institute, University of Warwick,
Coventry, CV4 7AL. UK.
j.c.robinson@warwick.ac.uk

Marco Romito*
Dipartimento di Matematica,
Università di Pisa,
Largo Bruno Pontecorvo 5,
56127 Pisa, Italia.
romito@dm.unipi.it

Edriss S. Titi*
Department of Mathematics,
Texas A&M University,
3368 TAMU, College Station, TX 77843-3368. USA.
titi@math.tamu.edu
&
Department of Computer Science and Applied Mathematics,
Weizmann Institute of Science,
Rehovot 76100. Israel.
edriss.titi@weizmann.ac.il

Alejandro Vidal-López*
Department of Mathematical Sciences,
Xi'an Jiaotong-Liverpool University,
Suzhou 215123, China P. R.
Alejandro.Vidal@xjtlu.edu.cn

Fei Wang
Department of Mathematics,
University of Southern California,
Los Angeles, CA 90089. USA.
wang828@usc.edu

Mohammed Ziane*
Department of Mathematics,
University of Southern California,
Los Angeles, CA 90089. USA.
ziane@math.usc.edu

1

Classical solutions to the two-dimensional Euler equations and elliptic boundary value problems, an overview

Hugo Beirão da Veiga

Dipartimento di Matematica,
Università di Pisa,
Via F. Buonarroti 1/c, Pisa. Italy.
bveiga@dma.unipi.it

Abstract

Consider the classical initial, boundary-value problem for the 2D Euler equations, which describes the motion of an ideal, incompressible, fluid in a impermeable vessel. In the early eighties we introduced and studied a Banach space, denoted $C_*(\overline{\Omega})$, which enjoys the following property: if the curl of the initial velocity belongs to $C_*(\overline{\Omega})$, and the curl of the external forces is integrable in time with values in the above space $C_*(\overline{\Omega})$, then all derivatives appearing in the differential equations and in the boundary conditions are continuous in space-time, up to the boundary (we call these solutions *classical solutions*). At that time this conclusion was know if $C_*(\overline{\Omega})$ is replaced by a Hölder space $C^{0,\lambda}(\overline{\Omega})$. In the proof of the above result we appealed to a $C^2(\overline{\Omega})$ regularity result for solutions to the Poisson equation, vanishing on the boundary and with external forces in $C_*(\overline{\Omega})$. Actually, at that time, we have proved this regularity result for solutions to more general second-order linear elliptic boundary-value problems. However the proof remained unpublished. Recently, we have published an adaptation of the proof to solutions of the Stokes system. We recall these results in Section 1.1 below. On the other hand, attempts to prove the above regularity results for data in functional spaces properly containing $C_*(\overline{\Omega})$, have also been done. Bellow we prove some partial results in this direction. This possibly unfinished picture leads to interesting open problems.

Published in *Recent Progress in the Theory of the Euler and Navier-Stokes Equations*, edited by James C. Robinson, José L. Rodrigo, Witold Sadowski, & Alejandro Vidal-López. ©Cambridge University Press 2016.

1.1 The Euler and Stokes equations with data in $C_*(\overline{\Omega})$.

In these notes we want to give an overview on some results, both old and new. Some are old, but remained unpublished for a long time. The starting point will be Beirão da Veiga (1981, 1982, 1984).

We start by introducing some notation. Ω is an open, bounded, connected set in $\mathbb{R}^n$, $n \geq 2$, locally situated on one side of its boundary Γ. We assume that Γ is of class $C^{2,\lambda}(\overline{\Omega})$, for some positive λ. By $C(\overline{\Omega})$ we denote the Banach space of all real, continuous functions in $\overline{\Omega}$ with the norm

$$\|f\| \equiv \sup_{x \in \overline{\Omega}} |f(x)|.$$

In the sequel we use the notation

$$\|\nabla u\| = \sum_{i=1}^{n} \|\partial_i u\|, \quad \|\nabla^2 u\| = \sum_{i,j=1}^{n} \|\partial_{ij} u\|,$$

and appeal to the canonical spaces $C^1(\overline{\Omega})$ and $C^2(\overline{\Omega})$, with the norms

$$\|u\|_1 \equiv \|u\| + \|\nabla u\|, \quad \|u\|_2 \equiv \|u\| + \|\nabla^2 u\|$$

respectively. Further, for each $\lambda \in (0,1]$, we define the semi-norm

$$[f]_{0,\lambda} \equiv \sup_{x,y \in \Omega; x \neq y} \frac{|f(x) - f(y)|}{|x - y|^\lambda}, \tag{1.1}$$

and the Hölder space $C^{0,\lambda}(\overline{\Omega}) \equiv \{f \in C(\overline{\Omega}) : [f]_{0,\lambda} < \infty\}$, with the norm

$$\|f\|_{0,\lambda} = \|f\| + [f]_{0,\lambda}.$$

In particular, $C^{0,1}(\overline{\Omega})$ is the space of Lipschitz continuous functions in $\overline{\Omega}$. By $C^\infty(\overline{\Omega})$ we denote the set of all restrictions to $\overline{\Omega}$ of infinitely differentiable functions in $\mathbb{R}^n$. We will use boldface notation to denote vectors, vector spaces, and so on. We denote the components of a generic vector u by u_i, and similarly for tensors. Norms in functional spaces whose elements are vector fields are defined in the usual way, by appealing to the corresponding norms of the components.

In considering the two-dimensional Euler equations we will introduce the following well-known simplification. For a scalar function $u(x)$ (identified here with the third component of a vector field, normal to the plane of motion) we define the vector field $\text{Rot}\,u = (\partial_2 u, -\partial_1 u)$. For a vector field $v = (v_1, v_2)$ we define the scalar field $\text{rot}\,v = \partial_1 v_2 - \partial_2 v_1$ (the normal component of the curl). One has $-\Delta = \text{rotRot}$. Note that

Rot$u$ is the rotation of the gradient ∇u by $\pi/2$ in the counter-clockwise direction.

Next we describe the motivation and origin of this research. We follow Beirão da Veiga (1981, 1982, 1984) which were essentially written during a visiting professorship to the Mathematics Research Center and the Mathematics Department in Wisconsin-Madison, in the semester October 1981-March 1982. In the above references we consider the initial boundary value problem for the two dimensional Euler equations

$$
\begin{cases}
\partial_t \boldsymbol{v} + (\boldsymbol{v} \cdot \nabla)\boldsymbol{v} = \boldsymbol{v} - \nabla\pi & \text{in} \quad Q \equiv \mathbb{R} \times \Omega, \\
\operatorname{div} \boldsymbol{v} = 0 & \text{in} \quad Q, \\
\boldsymbol{v}_0 \cdot \boldsymbol{n} = 0 & \text{on} \quad \mathbb{R} \times \Gamma, \\
\boldsymbol{v}(0) = \boldsymbol{v}_0 & \text{in} \quad \Omega.
\end{cases}
\tag{1.2}
$$

At that time our main interest was to determine minimal conditions on the data which imply that the global, unique, solutions to the above problem are *classical*. This means here that all derivatives appearing in the equations are continuous, up to the boundary, in the space-time cylinder. The main result on this problem was stated and proved in the preprint by Beirão da Veiga (1982), see the theorem 1.9 below. Exactly the same work was published in Beirão da Veiga (1984), to which we will refer in the sequel. To explain, in the simplest way, the main lines followed in our study, assume for now that no external forces are present, and that Ω is simply connected. In Beirão da Veiga (1984) we started by considering the Banach space

$$
\boldsymbol{E}(\overline{\Omega}) \equiv \{\boldsymbol{v} \in \mathbf{C}(\overline{\Omega}) : \operatorname{div}\boldsymbol{v} = 0 \text{ in } \Omega; \operatorname{rot}\boldsymbol{v} \in C(\overline{\Omega}); \boldsymbol{v} \cdot \boldsymbol{n} = 0 \text{ on } \Gamma\}, \tag{1.3}
$$

endowed with the norm (in the simply connected case)

$$
\|\|\boldsymbol{v}\|\| = \|\operatorname{rot}\boldsymbol{v}\|, \tag{1.4}
$$

and show the global boundedness, strong-continuous dependence on the data, and other basic properties with respect to data in the above space $\boldsymbol{E}(\overline{\Omega})$ (see the theorems 1.1, 1.2, and 1.3, in the above reference). These preliminary results were obtained by improving techniques already used by other authors; see for instance Kato (1967), and Schaeffer (1937). However these results do not imply that solutions are classical under the given assumption on the initial data, since

$$
\operatorname{rot}\boldsymbol{v}_0 \in C(\overline{\Omega})
$$

leads to $\operatorname{rot}\boldsymbol{v}(t, \cdot) \in C(\overline{\Omega})$, but this last property does not imply $\nabla\boldsymbol{v}(t, \cdot) \in$

$\mathbf{C}(\overline{\Omega})$. This gap is strictly related to a corresponding gap for solutions to elliptic equations, namely, the solution v to the system (see equation (1.3) in Beirão da Veiga, 1982)

$$\begin{cases} \operatorname{rot} v = f & \text{in } \Omega, \\ \operatorname{div} v = 0 & \text{in } \Omega, \\ v \cdot n = 0 & \text{on } \Gamma, \end{cases} \tag{1.5}$$

does not necessarily belong to $\mathbf{C}^1(\overline{\Omega})$, whenever $f \in C(\overline{\Omega})$. On the other hand, at that time, it was already well known that if f belongs to a Hölder space $C^{0,\lambda}(\overline{\Omega})$, then $v \in \mathbf{C}^{1,\lambda}(\overline{\Omega})$. This result, together with a clever use of Lagrangian coordinates, makes it possible to prove that solutions to the system (1.2) are classical under the hypothesis

$$\operatorname{rot} v_0 \in C^{0,\lambda}(\overline{\Omega}).$$

This was a well known result at that time, see Bardos (1972), Judovich (1963), Kato (1967), and Schaeffer (1937).

Having the above picture in mind, it seemed natural to start our approach to the Euler equations by studying the system (1.5). We wanted to single out a Banach spaces $C_*(\overline{\Omega})$, strictly contained in the Hölder spaces $C^{0,\lambda}(\overline{\Omega})$, such that solutions v to the first order system (1.5) are classical under the assumption $f \in C_*(\overline{\Omega})$. On the other hand, a classical argument shows that the solution v to the system (1.5) can be obtained by setting $v = -\operatorname{Rot} u$, where u solves the problem

$$\begin{cases} -\Delta u = f & \text{in } \Omega, \\ u = 0 & \text{on } \Gamma. \end{cases} \tag{1.6}$$

It follows that solutions v to system (1.5) belong to $\mathbf{C}^1(\overline{\Omega})$ if the solutions u to the system (1.6) belong to $\mathbf{C}^2(\overline{\Omega})$. This situation led us to look for a Banach space $C_*(\overline{\Omega})$, for which the following result holds.

Theorem 1.1.1. *Let $f \in C_*(\overline{\Omega})$ and let u be the solution to problem (1.6). Then $u \in C^2(\overline{\Omega})$, moreover, $\|u\|_2 \leq c_0\|f\|_*$.*

The above theorem was stated in Beirão da Veiga (1984) as Theorem 4.5. For convenience, the space $C_*(\overline{\Omega})$ will be defined at the end of this section.

Having obtained the above result, we succeeded in proving that the solutions to the Euler equations (1.2) are classical under the assumption

$$\operatorname{rot} v_0 \in C_*(\overline{\Omega}).$$

This is the main result in Beirão da Veiga (1984). More precisely, we proved the following statement.

Theorem 1.1.2. *Let* $\mathrm{rot}u_0 \in C_*(\overline{\Omega})$ *and* $\mathrm{rot}v \in L^1(\mathbb{R}^+; C_*(\overline{\Omega}))$. *Then, the global solution* v *to problem* (1.2) *is continuous in time with values in* $\mathbf{C}^1(\overline{\Omega})$, *that is*

$$v \in C(\mathbb{R}^+; \mathbf{C}^1(\overline{\Omega})). \tag{1.7}$$

Furthermore, the estimate

$$\|v(t)\|_{\mathbf{C}^1(\overline{\Omega})} \le c e^{c_1 B_t t}\{\|\mathrm{rot}v_0\|_{C_*(\overline{\Omega})} + \|\mathrm{rot}v\|_{L^1(0,t;C_*(\overline{\Omega}))}\} \tag{1.8}$$

holds for all $t \in \mathbb{R}^+$, *where*

$$B_t = \|\mathrm{rot}v_0\| + \|\mathrm{rot}v\|_{L^1(0,t;C(\overline{\Omega}))}. \tag{1.9}$$

Moreover, $\partial_t v$ *and* $\nabla\pi$ *are continuous in* $\overline{Q}$ *if both terms* v_0 *and* ∇F, *in the canonical Helmholtz decomposition* $v = v_0 + \nabla F$ *separately satisfy this same continuity property. Then all derivatives that appear in equations* (1.2) *are continuous in* $\overline{Q}$, *that is, we have a classical solution.*

The conclusion of the theorem is false in general for data $v_0 \in \mathbf{C}^1(\overline{\Omega})$, or $v \in L^1(\mathbb{R}^+; \mathbf{C}^1(\overline{\Omega}))$.

If Ω is not simply connected the results still apply, as remarked in Beirão da Veiga (1984), by appealing to well known devices. See, for instance, the appendix 1 in the above reference.

Concerning the 2D Euler equations, we also refer the reader to Koch (2002). In this interesting work the author considers not only the 2D Euler equations but also many other central problems. However, the claims and proofs that followed to treat the particular two-dimensional problem considered in reference Beirão da Veiga (1984) are not very dissimilar to those previously showed by us in this last reference. Related results can also be found in reference Vishik (1998).

In Beirão da Veiga (1984) it was remarked that Theorem 1.1.1 could also be extended to solutions to more general linear elliptic boundary value problems. In fact, in Beirão da Veiga (1981) we proved the following regularity result.

Theorem 1.1.3. *For every* $f \in C_*(\overline{\Omega})$ *the solution* u *to the problem*

$$\begin{cases} \mathcal{L}u = f & in \quad \Omega, \\ \mathcal{B}u = 0 & on \quad \Gamma, \end{cases} \tag{1.10}$$

belongs to $C^2(\overline{\Omega})$. Moreover, there is a constant c_0 such that the estimate

$$\|u\|_2 \leq c_0\|f\|_*, \quad \forall f \in C_*(\overline{\Omega}). \tag{1.11}$$

holds.

In the above theorem $\mathcal{L}$ is a second order partial differential elliptic operator with smooth coefficients, and $\mathcal{B}$ is a linear differential operator, of order less or equal to one, acting on the boundary Γ. In Beirão da Veiga (1981) we assumed that $\mathcal{L}$, $\mathcal{B}$, and Ω are such that, for each $f \in C(\overline{\Omega})$, problem (1.10) has a unique solution $u \in C^1(\overline{\Omega})$, given by

$$u(x) = \int_\Omega g(x,y)f(y)\,dy, \tag{1.12}$$

where g is the Green function associated with he above boundary value problem. Our hypotheses on $\mathcal{L}$, $\mathcal{B}$, and Ω are given by the following two requirements:

– For each $f \in C(\overline{\Omega})$ the solution u of problem (1.10) is unique, belongs to $C^1(\overline{\Omega})$, and is is given by (1.12). Furthermore, if $f \in C^\infty(\overline{\Omega})$ then $u \in C^2(\overline{\Omega})$.

– The above Green's function $g(x,y)$ satisfies the estimates

$$\left|\frac{\partial g}{\partial x_i}\right| \leq \frac{k}{|x-y|^{n-1}}, \quad \left|\frac{\partial^2 g}{\partial x_i \partial x_j}\right| \leq \frac{k}{|x-y|^n}, \tag{1.13}$$

where $i,j = 1,...,n$.

The above estimates for Green's functions have been well known for a large class of problems for a long time. Classical works are due to, for instance, Levi (1908, 1909), Hadamard (1914), Lichtenstein (1918), Eidus (1958), Levy (1920), and many other authors. We refer in particular to (Miranda, 1955, Chap. III, Sections 21, 22, and 23), and references therein (in particular, to Giraud's references). For much more general results on Green functions see Solonnikov (1970, 1971).

It is worth noting that the proof of Theorem 1.1.3 may be extended to a larger class of problems, like non-homogeneous boundary-value problems, elliptic systems, and in particular the Stokes system, higher order problems, etc. The main point is that solutions u are given by expressions like (1.12), where the Green's functions g satisfy suitable estimates, which extend that shown in equation (1.13). Recently, we have adapted the unpublished proof of theorem 1.1.3 to show a similar regularity result for solutions to the Stokes system (1.10). Actually, in Beirão da Veiga (2014) we prove the following result.

Theorem 1.1.4. *For every $f \in C_*(\overline{\Omega})$ the solution (u, p) of the Stokes system*

$$\begin{cases} -\Delta u + \nabla p = f & in \, \Omega, \\ \nabla \cdot u = 0 & in \, \Omega, \\ u = 0 & on \, \Gamma, \end{cases} \qquad (1.14)$$

belongs to $C^2(\overline{\Omega}) \times C^1(\overline{\Omega})$. Moreover, there exists a constant c_0, depending only on Ω, such that the estimate

$$\|u\|_2 + \|\nabla p\| \le c_0 \|f\|_*, \quad \forall f \in C_*(\overline{\Omega}), \qquad (1.15)$$

holds.

In the final part of the section we define the Banach space $C_*(\overline{\Omega})$. If $f \in C(\overline{\Omega})$ set, for each $r > 0$,

$$\omega_f(r) \equiv \sup_{x,y \in \Omega; 0 < |x-y| \le r} |f(x) - f(y)|, \qquad (1.16)$$

and define the semi-norm

$$[f]_* = [f]_{*,\delta} \equiv \int_0^\delta \omega_f(r) \frac{\mathrm{d}r}{r}. \qquad (1.17)$$

If $0 < \delta < R$, one has

$$[f]_{*,\delta} \le [f]_{*,R} \le [f]_{*,\delta} + 2\left(\log \frac{R}{\delta} \right) \|f\|. \qquad (1.18)$$

It follows that norms (obtained by the addition of $\|f\|$, see (1.20) below), are equivalent.

In the literature, the condition

$$\int_0^\delta \omega_f(r) \frac{\mathrm{d}r}{r} < +\infty$$

is called *Dini's continuity condition*, see Gilbarg & Trudinger (1977), equation (4.47). In Gilbarg & Trudinger (1977), problem 4.2, it is remarked that if f satisfies Dini's condition in $\mathbb{R}^n$, then its Newtonian potential is a C^2 solution of Poisson's equation $\Delta u = f$ in $\mathbb{R}^n$.

Definition 1.1.5.

$$C_*(\overline{\Omega}) \equiv \{f \in C(\overline{\Omega}) : [f]_* < \infty\}. \qquad (1.19)$$

As claimed in Beirão da Veiga (1984), $C_*(\overline{\Omega})$ endowed with the norm

$$\|f\|_* \equiv [f]_* + \|f\|, \tag{1.20}$$

is a Banach space, compactly embedded in $C(\overline{\Omega})$. Furthermore, $C^\infty(\overline{\Omega})$ is dense in $C_*(\overline{\Omega})$. We have appealed to these properties in reference Beirão da Veiga (1984), however the complete proofs were written only in an unpublished manuscript Beirão da Veiga (1981). For the complete proofs see the recent publication Beirão da Veiga (2014).

In Beirão da Veiga (1981) we introduced a functional space $B_*(\overline{\Omega})$, which strictly contains $C_*(\overline{\Omega})$, for which we have proven that the second order derivatives of the solutions to the system (1.10) are bounded in Ω for all $f \in B_*(\overline{\Omega})$. However, we did not succeed in proving, or disproving, the full result, namely, the continuity up to the boundary of the second order derivatives. This led us to leave unpublished the statements concerning the space $B_*(\overline{\Omega})$. In the next sections we show some of these results and proofs, and related open problems. Some results are proved below for data in a larger space $D_*(\overline{\Omega}) \supset B_*(\overline{\Omega})$.

As remarked in Beirão da Veiga (2014), another significant candidate could be obtained by replacing in the definition of $C_*(\overline{\Omega})$ given in (1.17) by the quantity $\omega_f(x; r)$ by

$$\widetilde{\omega}_f(x; r) = \sup_{x \in \Omega} \left| f(x) - |\Omega(x; r)|^{-1} \int_{\Omega(x;r)} f(y)\, \mathrm{d}y \right|. \tag{1.21}$$

1.2 The functional spaces $B_*(\overline{\Omega})$ and $D_*(\overline{\Omega})$.

In this section we define the spaces $B_*(\overline{\Omega})$ and $D_*(\overline{\Omega})$. We start with $B_*(\overline{\Omega})$. Set

$$\omega_f(x; r) = \sup_{y \in \Omega(x;r)} |f(x) - f(y)|, \tag{1.22}$$

and define, for each $x \in \overline{\Omega}$, the "point-wise" semi-norms

$$p_x(f) \equiv \int_0^\delta \omega_f(x; r) \frac{\mathrm{d}r}{r}, \tag{1.23}$$

and also the "global" semi-norm

$$\langle f \rangle_* = \sup_{x \in \overline{\Omega}} \int_0^\delta \omega_f(x; r) \frac{\mathrm{d}r}{r} = \sup_{x \in \overline{\Omega}} p_x(f). \tag{1.24}$$

Note that

$$[f]_* = \int_0^\delta \sup_{x \in \overline{\Omega}} \omega_f(x; r) \frac{dr}{r}. \tag{1.25}$$

Definition 1.2.1.

$$B_*(\overline{\Omega}) \equiv \{f \in C(\overline{\Omega}) : \langle f \rangle_* < +\infty\}. \tag{1.26}$$

The space $B_*(\overline{\Omega})$ endowed with

$$\|f\|^* \equiv \|f\| + \langle f \rangle_*, \tag{1.27}$$

is a normed linear space. Clearly $\langle f \rangle_* \leq [f]_*$. Further, in Beirão da Veiga (1981), we proved that the embedding $B_*(\overline{\Omega}) \subset C_*(\overline{\Omega})$ is strict, by constructing an oscillating function which belongs to $B_*(\overline{\Omega})$ but not to $C_*(\overline{\Omega})$; for the counterexample we take $\overline{\Omega} = [0, 1]$. We show this construction in Section 1.7 below.

Next we define $D_*(\overline{\Omega})$. Set

$$S(x; r) = \{y \in \Omega : |x - y| = r\},$$

and define

$$\mu_f(x; r) = \sup_{y \in S(x;r)} |f(x) - f(y)|, \tag{1.28}$$

for each fixed $x \in \overline{\Omega}$ and $r > 0$. Further, fix a real positive δ, and define the semi-norms

$$q_x(f) \equiv \int_0^\delta \mu_f(x; r) \frac{dr}{r}, \tag{1.29}$$

for each $x \in \overline{\Omega}$. As in (1.18), the particular positive value δ is not significant here. Note that the continuity of f at single point x follows necessarily from the finiteness of the integral in equation (1.29). To avoid unnecessary complications, we assume in the sequel that $f \in C(\overline{\Omega})$. Next define the semi-norm

$$(f)_* = \sup_{x \in \overline{\Omega}} \int_0^\delta \omega_f(x; r) \frac{dr}{r} = \sup_{x \in \overline{\Omega}} q_x(f). \tag{1.30}$$

It is worth noting that all the semi-norms introduced above enjoy property (1.18).

Definition 1.2.2.

$$D_*(\overline{\Omega}) \equiv \{f \in C(\overline{\Omega}) : (f)_* < +\infty\}. \qquad (1.31)$$

The linear space $D_*(\overline{\Omega})$ endowed with

$$\|\|f\|\|^* \equiv \|f\| + (f)_*, \qquad (1.32)$$

is a normed linear space. Obviously, $B_*(\overline{\Omega}) \subset D_*(\overline{\Omega})$. Finally, note that (1.18) holds for the above two functional spaces, with the obvious modifications.

1.3 Results and open problems.

Theorem 1.3.1. *Let $f \in D_*(\overline{\Omega})$, and let u be the solution to problem (1.10). Then the first order derivatives of the solution u are Lipschitz continuous in $\overline{\Omega}$. Furthermore, the estimate*

$$\|\nabla^2 u\|_{L^\infty(\Omega)} \leq c_0 \, \|\|f\|\|^* \qquad (1.33)$$

holds.

The proof of this result is an extension of the unpublished proof given in Beirão da Veiga (1981) for data $f \in B_*(\overline{\Omega})$. The proof will be shown in Section 1.4.

It remains an open problem whether the Theorems 1.1.3 and 1.1.4 hold with $C_*(\overline{\Omega})$ replaced by $B_*(\overline{\Omega})$ or by $D_*(\overline{\Omega})$. Let us discuss this point. Below we prove the following *conditional* result.

Theorem 1.3.2. *Let u be the solution of problem (1.10) with a given data $f \in D_*(\overline{\Omega})$. Assume that there is a sequence of data $f_m \in D_*(\overline{\Omega})$, convergent to f in $D_*(\overline{\Omega})$, such that the solutions u^m of problem (1.10) with data f_m belong to $C^2(\overline{\Omega})$. Then $u \in C^2(\overline{\Omega})$, and moreover*

$$\|\nabla^2 u\| \leq c_0 \|\|f\|\|^*. \qquad (1.34)$$

Theorem 1.3.2 will be proven in Section 1.5. It is worth noting that, since $B_*(\overline{\Omega}) \subset D_*(\overline{\Omega})$, the above two theorems hold with $D_*(\overline{\Omega})$ replaced by $B_*(\overline{\Omega})$, and $\|\|f\|\|^*$ replaced by $\|f\|^*$.

Corollary 1.3.3. *If $C^\infty(\overline{\Omega})$ is dense in $B_*(\overline{\Omega})$, then solutions u with data $f \in B_*(\overline{\Omega})$, belong to $C^2(\overline{\Omega})$, moreover*

$$\|u\|_2 \leq c_0 \|f\|^*, \quad \forall f \in B_*(\overline{\Omega}). \tag{1.35}$$

The result holds with $B_*(\overline{\Omega})$ replaced by $D_*(\overline{\Omega})$, and $\|f\|^*$ replaced by $\|\|f\|\|^*$.

In the above corollary we may replace $C^\infty(\overline{\Omega})$ by $C^{0,\lambda}(\overline{\Omega})$ (or even by $C_*(\overline{\Omega})$, as a consequence of Theorem 1.1.3). However we put here $C^\infty(\overline{\Omega})$ functions since there is a well known "two steps" argument used to prove density of $C^\infty(\overline{\Omega})$ in larger functional spaces. The first step consists of constructing a linear continuous map $f \to \widetilde{f}$, from $B_*(\overline{\Omega})$ to $B_*(\Omega_\delta)$, where

$$\Omega_\delta \equiv \{x : \mathrm{dist}(x, \Omega) < \delta\}, \tag{1.36}$$

for some $\delta > 0$, such that the restriction of $\widetilde{f}$ to $\overline{\Omega}$ coincides with f. In the second step, we appeal to the usual mollification technique to prove the desired density result in compact subsets of Ω_δ, so in $\overline{\Omega}$. The extension step is necessary, since approximation by mollification may hold only in compact subsets.

Concerning the first step, we prove the following "extension" result. For the proof, see the Section 1.6 below.

Proposition 1.3.4. *There exists $\delta > 0$, depending only on Ω, such that the following statement holds. There is a linear continuous map $f \to \widetilde{f}$, from $B_*(\overline{\Omega})$ to $B_*(\Omega_\delta)$, such that the restriction of $\widetilde{f}$ to $\overline{\Omega}$ coincides with f.*

Theorem 1.3.1, together with the ideas introduced by Beirão da Veiga (1984), provides new regularity results for solutions to the 2D Euler equations. This point will be considered in a forthcoming paper. For the time being, we merely state the following result.

Theorem 1.3.5. *Let $\mathrm{rot}\, v_0 \in B_*(\overline{\Omega})$, and assume that the external forces v vanish. Then, the global solution v to problem (1.2) satisfies*

$$\nabla u \in L^\infty(\overline{Q_T}), \tag{1.37}$$

for all $T > 0$.

The conclusion of the theorem is false in general for data $v_0 \in \mathbf{C}^1(\overline{\Omega})$.

It is worth noting that we have no reason to conjecture density of $C^\infty(\overline{\Omega})$ in $B_*(\overline{\Omega})$. In fact, our advise to readers interested in the subject is to try to prove the full Theorem 1.3.1 in the framework of the norm obtained by means of (1.21). In this context, it would be of interest to show that $C_*(\overline{\Omega})$ in strictly contained in the above new space.

1.4 Proof of Theorem 1.3.1.

We start by estimating the differential quotients of the first order derivatives of the solutions u.

Proposition 1.4.1. *Let $f \in C(\overline{\Omega})$, and let u be the solution to problem (1.10). Assume that for every given $x_0 \in \overline{\Omega}$, there exists a real $\delta_0 > 0$ such that*

$$q_{x_0}(f) \equiv \int_0^{\delta_0} \mu_f(x_0; r) \frac{dr}{r} < \infty. \tag{1.38}$$

Then

$$\left| \frac{\partial_i u(x) - \partial_i u(x_0)}{x - x_0} \right| \le c(q_{x_0}(f) + \|f\|), \tag{1.39}$$

for all $x \in \frac{\delta_0}{2}$, and $i = 1, ..., n$.

Proof. Let us introduce the auxiliary function $v(x)$ defined by

$$\begin{cases} \mathcal{L}v = 1 & \text{in} \quad \Omega, \\ \mathcal{B}v = 0 & \text{on} \quad \Gamma. \end{cases} \tag{1.40}$$

In particular $v \in C^{1,1}(\overline{\Omega})$, by assumption (i). Define

$$k_1 \equiv \|v\|_{1,1}. \tag{1.41}$$

Clearly k_1 depends only on $\mathcal{L}$, $\mathcal{B}$, and Ω, since v is completely determined by these data. Actually, k_1 depends only on some parameters related to the above elements (like the ellipticity constant of $\mathcal{L}$, for instance).

Define, in Ω, the functions $v^0(x) \equiv f(x_0)v(x)$, and $w(x) \equiv u(x) - v^0(x)$. Clearly, for each index $i = 1, ..., n$, we have

$$\partial_i w(x) = \int_\Omega \partial_i g(x, y)[f(y) - f(x_0)] \, dy, \quad \forall x \in \Omega.$$

Consequently,

$$|\partial_i w(x) - \partial_i w(x_0)| \le \int_\Omega |\partial_i g(x, y) - \partial_i g(x_0, y)||f(y) - f(x_0)| \, dy,$$

for all $x \in \Omega$.

Define

$$\Omega_c(x_0; r) = \Omega - \Omega(x_0; r),$$

and set $\rho = |x - x_0|$. We have

$$|\partial_i w(x) - \partial_i w(x_0)|$$

$$\leq \int_{\Omega(x_0; 2\rho)} |\partial_i g(x, y) - \partial_i g(x_0, y)||f(y) - f(x_0)|\, dy$$

$$+ \int_{\Omega(x_0; \delta_0) - \Omega(x_0; 2\rho)} |\partial_i g(x, y) - \partial_i g(x_0, y)||f(y) - f(x_0)|\, dy$$

$$+ \int_{\Omega_c(x_0; \delta_0)} |\partial_i g(x, y) - \partial_i g(x_0, y)||f(y) - f(x_0)|\, dy$$

$$\equiv I_1 + I_2 + I_3, \tag{1.42}$$

where $\delta_0 > 2\rho$.

Further, by appealing to the first estimate (1.13), we show that

$$I_1 \leq 2\|f\| \left\{ \int_{\Omega(x_0; 2\rho)} |\partial_i g(x_0, y)|\, dy + \int_{\Omega(x_0; 2\rho)} |\partial_i g(x, y)|\, dy \right\}$$

$$\leq c\|f\| \left\{ \int_{\Omega(x_0; 2\rho)} \frac{dy}{|x_0 - y|^{n-1}} + \int_{I(x; 3\rho)} \frac{dy}{|x - y|^{n-1}} \right\}. \tag{1.43}$$

Hence

$$I_1 \leq c\rho\|f\|.$$

On the other hand, by appealing to the mean-value theorem and the second estimate in (1.13), we find that

$$|\partial_i g(x, y) - \partial_i g(x_0, y)| \leq c\rho|x' - y|^{-n} \leq c\rho 2^n |x_0 - y|^{-n},$$

for each $y \in \Omega_c(x_0; 2\rho)$, where the point x' belongs to the straight seg-

H. Beirão da Veiga

ment joining x_0 to x. Consequently,

$$I_2 \leq c\rho \int_{\Omega(x_0;\delta_0)-\Omega(x_0;2\rho)} |f(y) - f(x_0)| \frac{\mathrm{d}y}{|x_0 - y|^n} \qquad (1.44)$$

$$\leq c\rho \int_{2\rho}^{\delta_0} \frac{\mathrm{d}r}{r^n} \int_{S(x_0;r)} \mu_f(x_0;r) \, \mathrm{d}S \qquad (1.45)$$

$$\leq c\rho S_n \int_{2\delta}^{\delta_0} \mu_f(x_0;r) \frac{\mathrm{d}r}{r}, \qquad (1.46)$$

where S_n is the area of the surface of the n-dimensional unit sphere. It follows that

$$I_2 \leq c\rho q_{x_0}(f).$$

Finally, a crude estimate for I_3 shows that

$$I_3 \leq c\rho \int_{\Omega_c(x_0;\delta_0)} \frac{1}{|x_0 - y|^n} |f(y) - f(x_0)| \, \mathrm{d}y \leq 2c\delta_0^{-n}\rho\|f\|.$$

By appealing to equation (1.42) and to the estimates proved above for I_1, I_2, and I_3, we obtain that

$$|\partial_i w(x) - \partial_i w(x_0)| \leq c\rho(q_{x_0}(f) + \|f\|).$$

Since

$$|\partial_i v^0(x) - \partial_i v^0(x_0)| \leq k_1\rho|f(x_0)|,$$

it follows that

$$c|\partial_i u(x) - \partial_i u(x_0)| \leq |\partial_i w(x) - \partial_i w(x_0)| + |\partial_i v^0(x) - \partial_i v^0(x_0)|$$

$$\leq c\rho\big(q_{x_0}(f) + \|f\| + k_1|f(x_0)|\big).$$

So,

$$\frac{|\partial_i u(x) - \partial_i u(x_0)|}{|x - x_0|} \leq c(q_{x_0}(f) + \|f\|), \quad \forall x \in \Omega, x \neq x_0. \qquad (1.47)$$

This shows (1.39), completing the proof.

$\square$

The proof of Theorem 1.3.1 follows immediately from proposition 1.4.1. Note that, by appealing to the first equation (1.13), we obtain

$$|\partial_i u(x)| \leq \|f\| \int_\Omega |\partial_i g(x,y)| \, \mathrm{d}y \leq c\|f\|, \quad \forall x \in \Omega.$$

Hence,

$$\|\nabla u\| \leq c\|f\|, \tag{1.48}$$

where u is the solution of problem (1.10).

1.5 Proof of Theorem 1.3.2.

Due to Theorem 1.3.1, it is sufficient to show that the second order classical derivatives of the solution $u(x)$ exist and are continuous, everywhere in $\overline{\Omega}$.

Consider the solutions u^m of problems

$$\begin{cases} \mathcal{L}u^m = f_m & \text{in} \quad \Omega, \\ \mathcal{B}u^m = 0 & \text{on} \quad \Gamma. \end{cases} \tag{1.49}$$

Clearly $u_m \to u$ in $C(\overline{\Omega})$. Further, by assumption, $u^m \in C^2(\overline{\Omega})$. By applying the result stated in Theorem 1.3.1 to the system

$$\begin{cases} \mathcal{L}(u^m - u^n) = f_m - f_n & \text{in} \quad \Omega, \\ \mathcal{B}(u^m - u^n) = 0 & \text{on} \quad \Gamma, \end{cases} \tag{1.50}$$

we obtain $\|\partial_{ij}u^m - \partial_{ij}u^n\| \leq c_0\|f_m - f_n\|^*$. This proves that $\partial_{ij}u^m$ is a Cauchy sequence in $C(\overline{\Omega})$. Hence, by the completeness of $C(\overline{\Omega})$, there exists an element $v_{ij} \in C(\overline{\Omega})$ such that the sequence $\partial_{ij}u^m$ is uniformly convergent in $\overline{\Omega}$ to v_{ij}. Furthermore, by applying estimate (1.48) to $u^m - u$, it follows that

$$\|\partial_i u^m - \partial_i u\| \leq c\|f_m - f\|.$$

Hence $\partial_i u^m$ converges uniformly in $\overline{\Omega}$ to $\partial_i u$.

The above results guarantee that the second order derivatives $\partial_{ij}u$ exist and are given by v_{ij}, for $i, j = 1, ..., n$.

1.6 Proof of Proposition 1.3.4.

In this section we prove the Proposition 1.3.4. The density of smooth functions is a crucial ingredient in proving full regularity. As already remarked, Proposition 1.3.4 is a typical first step to try to prove that smooth functions are dense in $B_*(\overline{\Omega})$. We start with some preliminary results. Recall that

$$\Omega_\delta \equiv \{x : \text{dist}(x, \Omega) < \delta\}.$$

It is well known that, for sufficiently small, positive δ, we can construct a suitable system of parallel surfaces Γ_r, where $-2\delta < r < 2\delta$, and $\Gamma_0 = \Gamma$. The surface Γ_r is at distance $|r|$ from Γ. It lies inside or outside Ω, depending on the negative or positive sign of the parameter r. Furthermore, a one to one correspondence between pairs of points in the opposite surfaces Γ_r and Γ_{-r} is defined by imposing that they belong to the same straight segment, orthogonal to Γ. We say that these points, denoted here by x and $\overline{x}$, are obtained from each other, by reflection (with respect to Γ).

Note that a positive lower bound for δ depends on the upper bound of the absolute values of the principal curvatures of the boundary Γ. We denote such a positive lower bound by δ, and use this same value in definitions (1.23) and (1.24).

Lemma 1.6.1. *There exist $\delta > 0$ and $k \geq 1$ (which depend only on the given set Ω), such that the following holds. Given $f \in C(\overline{\Omega})$, there is an extension $\tilde{f} : \Omega_{2\delta} \to \mathbb{R}$,*

$$\tilde{f}(x) = f(x) \quad \text{for} \quad x \in \overline{\Omega},$$

such that, for each $r \in (0, \delta)$,

$$\omega_{\tilde{f}}(x; r) \leq \begin{cases} \omega_f(x; kr), & \text{if } x \in \overline{\Omega}, \\ \omega_f(\overline{x}; kr), & \text{if } x \in \overline{\Omega}_\delta - \overline{\Omega}. \end{cases} \tag{1.51}$$

Proof. We define

$$\omega_f(x; r) = \sup_{y \in \Omega(x; r)} |f(x) - f(y)|, \tag{1.52}$$

and similarly,

$$\omega_{\tilde{f}}(x; r) = \sup_{y \in \Omega_{2\delta}(x; r)} |\tilde{f}(x) - \tilde{f}(y)|. \tag{1.53}$$

Obviously, if $x \in \Omega$, and $\text{dist}(x, \Gamma) \geq \delta$, it follows that

$$\omega_{\tilde{f}}(x; r) = \omega_f(x; r).$$

Hence we assume below that $\text{dist}(x, \Gamma) \leq \delta$.

In the sequel we show the basic ideas that lead to a more formal proof, whish is left to the interested reader.

We start by considering the particular case $x_0 \in \Gamma$, and by assuming that inside the δ-neighborhood of x_0 the boundary Γ is flat. Under these assumptions, compare the quantity

$$\omega_{\widetilde{f}}(x_0; r) = \sup_{y \in I(x_0; r)} |\widetilde{f}(x_0) - \widetilde{f}(y)|,$$

with that defined by (1.52) for $x = x_0$. In the case of $\widetilde{f}$, the points y describe a full ball, while, in the case of f, the points y describe a half ball. However, the set of numerical values $f(y)$ into play are, in both cases, exactly the same. The reader may verify that a similar situation occurs whenever the sphere $I(x; r)$ intersects Γ, in the flat case. The numerical values $f(y)$ into play are still the same, for $y \in I(x; r)$ and $y \in \overline{\Omega}(\overline{x}; r)$. The above remarks show that, in the locally-flat boundary case, (1.51) holds with $k = 1$.

In the general, non-flat, case, the geometrical situation is simply a deformation of the above one. Let us start by assuming that $x \notin \overline{\Omega}$, and $I(x, r)$ does not intersect Γ. Here, contrary to the flat boundary case, the reflection of $I(x, r)$ is not $I(\overline{x}, r)$. However, the reflection is contained in a (possibly large) sphere $I(\overline{x}, kr)$ (roughly speaking, $k \leq 1$, if Γ is locally convex, and $k \geq 1$, if Γ is locally concave). Since we assume that Ω is regular (in particular, locally situated in one side of the boundary) the local values obtained for k are uniformly bounded from above.

Finally, if $I(x, r)$ intersects the boundary Γ, the more general picture is simply an overlap of the two single situations, already described. Details are left to the reader.

$\square$

Proof of proposition 1.3.4. By Lemma 1.6.1, we may construct an extension $\widetilde{f} : \Omega_{2\delta} \to \mathbb{R}$ of the given function f, such that (1.51) holds. It follows that, for each $x \in \overline{\Omega}$,

$$\int_0^\delta \omega_{\widetilde{f}}(x; r) \frac{\mathrm{d}r}{r} \leq \int_0^\delta \omega_f(x; kr) \frac{\mathrm{d}r}{r},$$

and, for each $x \in \overline{\Omega}_\delta - \overline{\Omega}$,

$$\int_0^\delta \omega_{\widetilde{f}}(x; r) \frac{\mathrm{d}r}{r} \leq \int_0^\delta \omega_f(\overline{x}; kr) \frac{\mathrm{d}r}{r} = \int_0^{k\delta} \omega_f(\overline{x}; r) \frac{\mathrm{d}r}{r}.$$

In particular, it follows (note that extension of formula (1.18) holds) that

$$\langle \widetilde{f} \rangle_{*,\delta} \leq \langle f \rangle_{*,k\delta} \leq \langle f \rangle_{*,\delta} + 2 \Big(\log k \Big) \|f\|.$$

□

1.7 The embedding $B_*(\overline{\Omega}) \subset C_*(\overline{\Omega})$ is strict.

Below we construct an oscillating function in the interval $[0,1]$ which belongs to $B_*(\overline{\Omega})$ but not to $C_*(\overline{\Omega})$.

Proposition 1.7.1. *The inclusion $C_*(\overline{\Omega}) \subset B_*(\overline{\Omega})$ is proper.*

Proof. For each non-negative integer n set

$$r_n = e^{-2^n},$$

and define, in the interval $I = [0,1]$, the real continuous function

$$f_n(x) = \begin{cases} \frac{2^{-n}}{r_n}(x - (2^{-n} - r_n)), & \text{if } 2^{-n} - r_n \le x \le 2^{-n}; \\ \frac{2^{-n}}{r_n}((2^{-n} + r_n) - x), & \text{if } 2^{-n} \le x \le 2^{-n} + r_n; \qquad (1.54) \\ 0, & \text{if } |x - 2^{-n}| \ge r_n. \end{cases}$$

Note that f_n is linear in the intervals $[2^{-n} - r_n, 2^{-n}]$ and $[2^{-n}, 2^{-n} + r_n]$. Below we work with $\delta = r_0 = \frac{1}{e}$ in Definition 1.17.

In I, define the function

$$f(x) = \sum_{n=0}^{\infty} f_n(x).$$

Note that $0 \le f(x) \le x$. We start by showing that $f \notin C_*(\overline{\Omega})$. For convenience, we set

$$\omega(r) = \omega_f(r).$$

Clearly, if $r_{n+1} \le r \le r_n$ then

$$2^{-(n+1)} = \omega(r_{n+1}) \le \omega(r) \le \omega(r_n) = 2^{-n}.$$

Hence

$$\int_0^{\delta} \omega(r) \frac{dr}{r} \ge \sum_{n=0}^{\infty} \int_{r_{n+1}}^{r_n} \omega(r_{n+1}) \frac{dr}{r} \ge \sum_{n=0}^{\infty} \frac{1}{2} = \infty.$$

This shows that $f \notin C_*(\overline{\Omega})$.

Next we prove that $f \in B_*(\overline{\Omega})$. We want to show that there exists a constant c_0 such that, for all $x \in I$,

$$\int_0^{\delta} \omega_f(x; r) \frac{dr}{r} \le c_0. \qquad (1.55)$$

We start by considering the points $x_n = 2^{-n}$ where the function $f(x)$ attain local maximum values. Actually, $f(2^{-n}) = 2^{-n}$. Other points $x \in I$ can be treated by following a similar argument.

Let $x_n = 2^{-n}$ be fixed. One has

$$\omega_f(x_n; r) \leq \begin{cases} 2^{-n}\frac{r}{r_n}, & \text{if } 0 < r \leq r_n; \\[2mm] 2^{-n+1}, & \text{if } r_n \leq r \leq 2^{-n}; \\[2mm] 2^{-n} + r, & \text{if } 2^{-n} \leq r \leq r_0. \end{cases} \tag{1.56}$$

Note that equality holds in the first row of equation (1.56). Further, in the second row, the inequality $r \leq 2^{-n}$ should be interpreted as $x_n + r \leq 2^{-n+1}$. The second and third rows follow from the inequality

$$\omega(x, r) \leq x + r,$$

which holds for all $x \in I$ and $r > 0$.

Next, in accordance with (1.56), we decompose the integral on the left hand side of equation (1.55) as the sum of three integrals, and we estimate each of the integrals by appealing to the related region in (1.56). After some elementary calculations (left to the reader) it is east to see that equation (1.55) holds with $c_0 = 3$, for each point x_n. Finally note that, for $x = 0$, equation (1.55) holds with $c_0 = 1$.

Assume now that

$$x_{n-1} < x < x_n, \tag{1.57}$$

for some index n. We will follow the argument used above. As before,

$$\omega_f(x; r) \leq \begin{cases} 2^{-n+1}, & \text{if } r_n \leq r \leq 2^{-n}; \\[2mm] 2^{-n} + r, & \text{if } 2^{-n} \leq r \leq r_0. \end{cases} \tag{1.58}$$

It remains to consider the case in which $0 < r \leq r_n$. Under this assumption, and by taking into account (1.57), it readily follows that points y satisfying $|y - x| < r_n$ do not reach the supports of f_{n+2} and f_{n-1}. Hence we may replace here f by $f_n + f_{n+1}$. It follows that

$$\omega_f(x; r) \leq \omega_{(f_n + f_{n+1})}(x; r) \leq \omega_{f_n}(x; r) + \omega_{f_{n+1}}(x; r),$$

for $0 < r \leq r_n$. The above considerations lead to (1.55). $\qquad \square$

We remark that the above function f is not the limit in the $B_*(\overline{\Omega})$ norm of the sequence of Lipschitz continuous functions

$$F_N(x) = \sum_{n=1}^{N} f_n(x),$$

since $\langle f_n \rangle_* \geq \frac{1}{2}$, for every n. This fact does not exclude the possibility of approximating f by a sequence of elements belonging to $C_*(\overline{\Omega})$.

References

Bardos, C. (1972) Existence et unicité de la solution de l'équation de Euler en dimension deux. *J. Math. Anal. Appl.* **40**, 769–790.

Beirão da Veiga, H. (1981-82) unpublished manuscript.

Beirão da Veiga, H. (1982) *On the solutions in the large of the two-dimensional flow of a non-viscous incompressible fluid.* MRC Technical Summary Report no. 2424.

Beirão da Veiga, H. (1984) On the solutions in the large of the two-dimensional flow of a nonviscous incompressible fluid. *J. Diff. Eq.* **54**, no. 3, 373–389.

Beirão da Veiga, H. (2014) Concerning the existence of classical solutions to the Stokes system. On the minimal assumptions problem. *J. Math. Fluid Mech.* **16**, 539–550.

Eidus, D.M. (1958) Inequalities for Green's function. *Mat. Sb.* **45** (87), 455–470.

Gilbarg, D. & Trudinger, N.S. (1977) *Elliptic Partial Differential Equations of Second Order*, Springer-Verlag, Berlin/Heidelberg/New-York.

Hadamard, J. (1914) A propos d'une note de M. Paul Lévy sur la fonction de Green. *C. R. Acad. Sci. Paris* **158**, 1010–1011.

Judovich, V.I. (1963) Non-stationary flow of an ideal incompressible liquid. *Zh. Vychisl. Mat. i Mat. Fiz.* **3**, 1032–1066.

Kato, T. (1967) On classical solutions of the two-dimensional non-stationary Euler equation. *Arch. Rat. Mech. Anal.* **25**, 188–200.

Koch, H. (2002) Transport and instability for perfect fluids. *Math. Ann.* **323**, 491–523.

Levi, E.E. (1908) Sur l'application des équations intégrales au probleme de Riemann. *Nachrichten von der Kgl. Gesellschaft der Wissenschaften zu Göttingen*, 249–252.

Levi, E.E. (1909) I problemi dei valori al contorno per le equazioni lineari totalmente ellitiche alle derivate parziali, *Memorie della Società Italiana delle Scienze*.

Lévy, P. (1920) Sur l'allure des fonctions de Green et de Neumann dans le voisinage du contour. *Acta Math.* **42**, 207–267.

Lichtenstein, L. (1918) *Neure Entwicklung der Potentialtheorie, Konforme Abbildung.* Encycl. Math. Wissenschaft, Band II, 3. Teil, 1 Hefte; Leipzig, 7–248.

Miranda, C. (1955) *Equazioni alle Derivate Parziali di Tipo Ellittico.* Springer-Verlag, Berlin.

Schaeffer, A.C. (1937) Existence theorem for the flow of an ideal incompressible flow in two dimensions. *Trans. Amer. Math. Soc.* **42**, 497–513.

Solonnikov, V.A. (1970) On Green's Matrices for Elliptic Boundary Problem I. *Trudy Mat. Inst. Steklov* **110**, 123–170.

Solonnikov, V.A. (1971) On Green's Matrices for Elliptic Boundary Problem II. *Trudy Mat. Inst. Steklov* **116**, 187–226.

Vishik, M. (1998) Hydrodynamics in Besov spaces. *Arch. Rational Mech. Anal.* **145**, no. 4, 197–214.

2

Analyticity radii and the Navier–Stokes equations: recent results and applications

Zachary Bradshaw

Department of Mathematics,
University of British Columbia,
1984 Mathematics Road,
Vancouver, B.C.
V6T 1Z2. Canada.
`zbradshaw@math.ubc.ca`

Zoran Grujić

Department of Mathematics,
Kerchof Hall,
University of Virginia,
Charlottesville, VA 22904. USA.
`zg7c@virginia.edu`

Igor Kukavica

Department of Mathematics,
University of Southern California,
Los Angeles, CA 90089. USA.
`kukavica@usc.edu`

Abstract

This note highlights recent work involving the analyticity radii of solutions to the 3D Navier–Stokes equations. In particular it includes estimates for a solution's local analyticity radius at interior points of possibly bounded domains and, as an application, a conditional regularity criteria in weak-L^3.

2.1 Introduction

In this note we describe recent results involving the analyticity radii of solutions to the Navier–Stokes equations. The 3D Navier–Stokes equations govern the evolution of a viscous incompressible fluid's velocity

Published in *Recent Progress in the Theory of the Euler and Navier-Stokes Equations*, edited by James C. Robinson, José L. Rodrigo, Witold Sadowski, & Alejandro Vidal-López. ©Cambridge University Press 2016.

field u subjected to a forcing f and read

$$\partial_t u - \nu \Delta u = -u \cdot \nabla u - \nabla p + f \qquad \text{in } \Omega \times (0, T)$$
$$\nabla \cdot u = 0 \qquad\qquad\qquad\qquad\quad \text{in } \Omega \times (0, T) \qquad \text{(3D NSE)}$$
$$u(\cdot, 0) = u_0(\cdot) \qquad\qquad\qquad\quad \text{in } \Omega,$$

where the scalar valued function p represents the fluid's pressure, ν is the viscosity coefficient, u_0 is the initial data which is taken in an appropriate function space, and $\Omega \subseteq \mathbb{R}^3$ is a domain. When a boundary is present, Dirichlet boundary conditions are imposed along with a normalising condition on p.

A motivation for studying the analyticity radius of solutions to the 3D NSE lies in their connection to the dissipative scale in turbulence, see Foias (1995), Henshaw, Kreiss, & Reyna (1995, 1990), Monin & Yaglom (1971). The turbulent energy cascade refers to a process in which energy is transported by inertial mechanisms from larger to smaller scales in a statistically regular manner (see Frisch, 1995). In the absence of viscosity these dynamics saturate all scales below some macroscale associated with the fluid's environment and forcing. In viscous flows, friction dissipates energy at small scales which breaks down the Eulerian cascade dynamics. The dissipative scale refers to the length at which inertial forces become subordinate to the diffusive effects of viscosity. This is realized mathematically as the exponential decay of the Fourier spectrum of an analytic solution to 3D NSE at frequencies beyond the inverse of the analyticity radius.

The real analyticity of strong solutions to 3D NSE is a classical result (see Giga, 1983, Komatsu, 1979, Masuda, 1967). Early estimates from below for a solution's analyticity radius were provided in Foias & Temam (1989), Komatsu (1979). In Foias & Temam (1989), these estimates are obtained using Gevrey classes in an L^2 setting. As this approach depends heavily on Fourier analysis, it is most applicable to uniformly real analytic solutions and does not clearly extend to formulations of 3D NSE on domains possessing boundaries or where the forcing is only locally analytic. The approach of Komatsu (1979) applies to bounded domains but requires highly technical recursive estimates. Another approach for unbounded and periodic settings was developed in Grujić & Kukavica (1998) and can be applied to local contexts. Indeed, in Grujić & Kukavica (1999) estimates for the local analyticity radius are given for a solution to a nonlinear heat equation at interior points of a bounded domain. An analogous result for 3D NSE is complicated by the non-locality of the pressure. This is addressed in Bradshaw, Grujić,

& Kukavica (2015) where the authors provide local estimates for the analyticity radii of solutions to 3D NSE with possibly non-uniformly analytic forcing.

Analyticity radii are also connected to the regularity of certain solutions to 3D NSE. In Grujić (2013), a geometric measure-type conditional regularity criteria is given based on the sparseness of the superlevel sets of the vorticity magnitude. The condition is reasonable given the physically and numerically supported view that these sets have a filamentary geometry wherein individual vortex tubes are typically stretched in the axial direction, see Frisch (1995). Put roughly, the regularity criteria of Grujić (2013) is triggered whenever transverse length scales of vortex structures become smaller than a fraction of the vorticity's analyticity radius. In this sense the analyticity radius provides a critical length scale against which decaying filamentary radii can be compared. Subsequent work which identifies criticality breaking conditions has been carried out in Bradshaw & Grujić (2015) and also in Bradshaw & Grujić (2014).

The mathematical insight of Grujić (2013) also applies in an amorphous scenario and leads to a regularity criteria based on the smallness of a solution in the weak Lebesgue space L_w^3. Indeed, there exists a small universal constant $\epsilon_0 > 0$ so that, if

$$\sup_{0<t<T} \|u\|_{L_w^3(\mathbb{R}^3)} < \epsilon_0,$$

where u is a regular solution to 3D NSE on $(0, T)$, then u is regular at T. If the L_w^3 norm is replaced by the L^3 norm the smallness condition can be removed, as proved in the paper of Escauriaza, Seregin, & Šverák (2003). While an analogous result in L_w^3 remains a challenging open problem, a variety of partial results exist (see Kim & Kozono, 2006, 2004, Luo & Tsai, 2013). Our method is noteworthy for its simplicity and the fact that it does not depend on any Leray-type *a priori* bounds.

This note is structured as follows. Section 2.2 contains a brief survey of past results on the topic of analyticity radii in 3D NSE and related fluid models. In Section 2.3 we discuss recent work which provides lower bounds for analyticity radii in local contexts. Finally, in Section 2.4, we state and prove a conditional regularity criteria in L_w^3.

2.2 A review of known results

A function f defined on a domain $\Omega \subseteq \mathbb{R}^D$ is real analytic if, for any bounded subdomain $\Omega' \subseteq \Omega$, there exist positive numbers $M = M(\Omega', f)$

and $\lambda = \lambda(\Omega', f)$ so that

$$|\partial^\alpha f(x)| \le M \frac{\alpha!}{\lambda^{|\alpha|}},$$

for every $x \in \Omega'$ and multi-index $\alpha \in \mathbb{N}^D$. The constant λ is a lower bound on the analyticity radius of f on Ω'.

Early results on the analyticity of solutions to 3D NSE can be found in Kahane (1969), Masuda (1967). Masuda (1967) proved that sufficiently regular weak solutions to the 3D NSE no-slip boundary value problem with space- and time-analytic forcing are analytic and applied this to obtain a unique continuation result. Later, Kahane (1969) used representation formulas developed in Serrin (1962) to estimate successive spatial derivatives of weak solutions following a procedure for establishing space analyticity originally developed by Gevrey (1918). Both of these approaches established interior analyticity but did not extend to the boundary. This was addressed in Komatsu (1979) by estimating successive derivatives assuming the force is real analytic and also in Nakagawa (1981) under an additional assumption that the initial data is analytic using the classical approach of Morrey (1958).

Information about the analyticity radius is difficult to extract from the works listed above. In their pioneering work Foias and Temam (1989) used Hilbert space techniques to establish the time analyticity of solutions in Sobolev spaces of periodic functions. Analogous results for spatial analyticity are contained in Ferrari & Titi (1998) and Oliver & Titi (2001, 2000) along with general remarks about the approach. The novel aspect of Foias & Temam (1989) is the use of Gevrey class regularity. Gevrey classes are defined using the pseudo-differential operator $A = \sqrt{-\Delta}$ and its powers which are characterized through Fourier multipliers. Note that the norm of the Sobolev space $H^r(\mathbb{R}^3)$ is equivalent to $\| \cdot \|_{L^2(\mathbb{R}^3)} + \|A^r \cdot \|_{L^2(\mathbb{R}^3)}$. For $\tau, r > 0$ the Gevrey class r is the space

$$D(\mathrm{e}^{\tau A} : H^r) = \{f \in H^r(\mathbb{R}^D) : \mathrm{e}^{\tau A} f \in H^r(\mathbb{R}^D)\},$$

which is endowed with the norm

$$\| \cdot \|_{L^2(\mathbb{R}^3)} + \|A^r \mathrm{e}^{\tau A} \cdot \|_{L^2(\mathbb{R}^3)}.$$

On $\mathbb{R}^3$, elements of $D(\mathrm{e}^{\tau A} : H^r)$ are real analytic and posses analyticity radii bounded below by $\tau/\sqrt{D}$.

Using these structures it is possible to show that sufficiently regular solution to 3D NSE are real analytic and that the domain of analyticity of their complexification includes

$$\{x + iy : x \in \Omega, \ |y| \lesssim \sqrt{t}\}.$$

A benefit of working with this formalism is that it eliminates the need to compute estimates of higher derivatives such as those in Kahane (1969) and Komatsu (1979).

The Gevrey space technique was extended from $L^2(\mathbb{R}^3)$ to $L^p(\mathbb{R}^3)$ by Lemarié-Rieusset (2000, 2002, 2004). Subsequent adaptations have been developed for initial data in l^p (see Biswas, 2005, Biswas & Swanson, 2007, Biswas et al., 2014) and critical Besov spaces (see Bae, Biswas, & Tadmor, 2012). The procedure also applies to a other fluid models, e.g. the primitive equations of the ocean (see Ignatova, Kukavica, & Ziane, 2012), second grade fluid equations (see Paciu & Vicol, 2011), the critical and subcritical surface quasi-geostrophic equations (see Bae, Biswas, & Tadmor, to appear), and the Euler equations (see Kukavica & Vicol, 2009, 2011, Levermore & Oliver, 1997).

An alternative approach for estimating analyticity radii was developed in Grujić & Kukavica (1998). This strategy is based on the fact that elements of a well-known approximation scheme, see Kato (1984), are real analytic and, for strong solutions in $L^p(\mathbb{R}^3)$ with $p > 3$, these analytical properties extend to the limiting solution. The approximation scheme is constructed by setting $u^{(0)} = p^{(0)} = 0$ and iteratively solving

$$
\begin{aligned}
\partial_t u^{(n)} - \Delta u^{(n)} &= -u^{(n-1)} \cdot \nabla u^{(n-1)} - \nabla p^{(n-1)} + f &&\text{in } (0, \infty) \times \mathbb{R}^3 \\
\nabla \cdot u^{(n)} &= 0 &&\text{in } (0, \infty) \times \mathbb{R}^3 \\
u^{(n)}(x, 0) &= u_0(x) &&\text{in } \mathbb{R}^3
\end{aligned}
$$

where $p \in (3, \infty]$, the initial data and force are in $L^p(\mathbb{R}^3)$, and f is uniformly real analytic with analyticity radius λ_f. The extension of f into $\mathbb{C}^3$ is denoted $F + iG$. Regularity results for the heat and Poisson equations imply that elements of this scheme inherit the analytical properties of f. Denote by $U^{(n)} + iV^{(n)}$ the extension of $u^{(n)}$ into $\mathbb{C}^3$ and by $U_\alpha^{(n)}$ and $V_\alpha^{(n)}$ the evaluation $U^{(n)}$ and $iV^{(n)}$ at $x + i\alpha t$ where $\alpha \in \mathbb{R}^3$. Provided $|\alpha|t \leq \lambda_f$, the Cauchy–Riemann system is satisfied by $U_\alpha^{(n)}$ and $V_\alpha^{(n)}$ and it is simple to derive a coupled parabolic system over $\mathbb{R}^3$ which governs their evolutions. Proving the convergence of these extensions to the extension of the strong solution on a complex domain surrounding $\mathbb{R}^3$ depends on providing uniform estimates in $L^\infty(0, T; L^p(\mathbb{R}^3))$ using a mild solution approach. The resulting theorem from Grujić & Kukavica (1998) is the following.

Theorem 2.2.1 (Grujić & Kukavica, 1998). *Assume that $\|u_0\|_{L^p} \leq M_p < \infty$ and suppose*

$$M_f = \sup_{t \geq 0} \sup_{|y| < \lambda_f} \left(\|F(\cdot + iy, t)\|_{L^p(\mathbb{R}^3)} + \|G(\cdot + iy, t)\|_{L^p(\mathbb{R}^3)} \right) < \infty,$$

where $p \in (3, \infty)$. Let

$$T = \min \left\{ \frac{1}{C\, p^2\, M_p^{2p/(p-3)}}, \frac{M_p}{C\, M_f} \right\}.$$

Then there exists a solution u to 3D NSE on $(0, T) \times \mathbb{R}^3$ which agrees with the restriction to $\mathbb{R}^3$ of function $u^{(n)} + iv^{(n)}$ of three complex variables which is analytic on

$$\left\{ x + iy \in \mathbb{C}^3 : |y| \leq C^{-1} \sqrt{t} \wedge \lambda_f \right\},$$

at times $t \in (0, T)$, where by $\wedge$ we denote minimum.

This technique has been applied to the vorticity in Kukavica (1999) and the limiting case $p = \infty$ in Guberović (2010). In Grujić & Kukavica (1999) this approach was refined to provide lower bounds on the analyticity radius of solutions to a nonlinear heat equation at interior points of a bounded domain Ω. As in Grujić & Kukavica (1998), the proof is based on the analytic extensions of elements of an approximation scheme but here the complexification is carried out locally by evaluating at complex points of the form $x + i\alpha\,\psi(x)\,t$ for t positive, $x \in \Omega$, and a nonnegative test function ψ compactly supported in Ω. Estimates on the analyticity radius are obtained by considering these extensions globally as solutions to second order parabolic initial-boundary value problems over Ω. As formulated in Grujić & Kukavica (1999), this technique does not extend to 3D NSE, a fact owing to the non-locality introduced by the pressure. This has led us to construct a new, purely local approach which is contained in Bradshaw, Grujić, & Kukavica (2015) and summarized in the next section of this note.

2.3 Local estimates on analyticity radii

In this section we highlight recent work by the authors which introduces a new method for establishing the local analyticity of a solution to 3D NSE and estimating its the local analyticity radius. This approach accommodates locally analytic forcing and is applicable to boundary value formulations of 3D NSE.

Fix a point x_* and a time T_0, and assume that for all $t \in (0, T_0)$ the force $f(\cdot, t)$ is real analytic at x_* with analyticity radius r_*. Denote by kB_* the ball centered at x_* of radius kr_*. It follows that f agrees with the restriction of an analytic function $F + iG$ to B_*. Moreover, $F + iG$ is complex analytic on the domain

$$\Omega_{f,T_0}(x_*) = \left\{ x + iy \in \mathbb{C}^3 : x \in B_*, |y| < r_* - |x - x_*| \right\} \times (0, T_0),$$

for all $t \in (0, T_0)$. Let $\psi \colon \mathbb{R}^3 \to [0, 1]$ be a radial test function supported on B_* which evaluates to 1 on $B_*/2$, is decreasing with respect to $|x_* - x|$, and moreover satisfies

$$\|\nabla \psi\|_\infty \leq \frac{C}{r^*}.$$

We denote by S_f the collection of vectors $\alpha \in \mathbb{R}^3$ satisfying

$$(x + i\alpha\, \psi(x)\, t, t) \in \Omega_{f,T_0}(x_*)$$

and let $F_\alpha(x, t) = F(x + i\alpha\, \psi(x)\, t, t)$ and $G_\alpha(x, t) = G(x + i\alpha\, \psi(x)\, t, t)$.

We work with a solution u of 3D NSE on a domain containing $2B_*$ and also assume that u is locally smooth on $(0, T_0) \times 2B_*$. Our result involves the two quantities

$$M_q = \sup_{0 < t < T_0} \|u(t)\|_{L^q(2B_*)} + \sup_{0 < t < T_0} \|p(t)\|^{1/2}_{L^{q/2}(2B_*)}$$

$$+ T_0^{(r-2)/(2r)} \left(\int_0^{T_0} \|\nabla u(t)\|^r_{L^q(2B_*)}\, dt \right)^{1/r},$$

and

$$M_f = \sup_{\alpha \in S_f} \left(\sup_{0 < t < T} \|F_\alpha(t)\|_{L^q(2B_*)} + \sup_{0 < t < T} \|G_\alpha(t)\|_{L^q(2B_*)} \right),$$

where $q > 3$ and $r > 2q/(q - 3)$. Provided these are finite we specify a time T_1 satisfying

$$T_1 \leq \frac{1}{C} \min \left\{ T_0, r_*^2, \frac{1}{q^2\, M_q^{2q/(q-3)}}, \frac{M_q}{M_f} \right\}.$$

We are now ready to state the main result of Bradshaw, Grujić, & Kukavica (2015).

Theorem 2.3.1. *Let u be a locally smooth solution to 3D NSE on $(0, T_0) \times 2B_*$ and assume M_q and M_f are finite. Then, at all times $t \in (0, T_1)$, u agrees with the restriction to B_* of a function $U(x, y, t) + iV(x, y, t)$ which is analytic in the region*

$$\Omega_*(t) = \left\{ x + iy \in \mathbb{C}^3 : x \in B_*, |y| < \frac{\sqrt{t}}{4\, C_0} \right\},$$

for a universal constant C_0.

Let $\lambda_*(t)$ denote the local analyticity radius of u at x_*. The prior theorem implies a lower bound on $\lambda_*(t)$ for times in $(0, T_1)$, namely,

$$\lambda_*(t) \ge C_* \max\left\{ r_*, \frac{1}{q\, M_q^{q/(q-3)}}, \frac{M_q}{M_f}\right\},$$

which is consistent with the global estimates found in Grujić & Kukavica (1998).

Proof. (sketch) In order to prove Theorem 2.3.1, we introduce a test function ϕ which is taken to be the dilation of ψ by a factor of 2. Since u is locally smooth, the product ϕu satisfies

$$\partial_t(\phi u) - \Delta(\phi u) = -(\phi u) \cdot \nabla(\phi u) - \nabla(\phi p) + \Phi_1 + \phi f,$$

where

$$\Phi_1 = \phi(1 - \phi)\nabla \cdot (u \otimes u) + (\nabla \phi \cdot u)\phi u - 2\nabla\phi \cdot \nabla u - 2\Delta\phi\, u + p\nabla\phi.$$

Also, ϕp satisfies

$$\Delta(\phi p) = -\partial_i\partial_j(\phi u_i \phi u_j) + \Phi_2,$$

where

$$\Phi_2 = (1 - \phi)\, \phi\, \partial_i\partial_j(u_i u_j) + \partial_i\partial_j(\phi^2)u_i u_j + 2\nabla\phi\nabla p + (\Delta\phi)\, p.$$

Letting $\mu_0 = \phi u_0$ and setting $\mu^{(0)} = \rho^{(0)} = 0$ we construct a sequence of approximations by solving

$$\partial_t\mu^{(n)} - \Delta\mu^{(n)} = -\mu^{(n-1)} \cdot \nabla\mu^{(n-1)} - \nabla\rho^{(n-1)} + \Phi_1 + \phi f$$
$$\Delta\rho^{(n-1)} = -\partial_i\partial_j\big(\mu_i^{(n-1)}\mu_j^{(n-1)}\big) + \Phi_2$$

in $\mathbb{R}^3 \times (0, T)$ and

$$\mu^{(n)}(\cdot, 0) = \mu_0(\cdot) \quad \text{in } \mathbb{R}^3.$$

By properties of the heat and Poisson equations as well as the analytical properties of f at x_*, the scheme elements $\mu^{(n)}$ and $\rho^{(n)}$ are real analytic at x_* with analyticity radii at least r_*. We denote their extensions by $U^{(n)} + iV^{(n)}$ and $P^{(n)} + i\Pi^{(n)}$. As above we use a subscript α to indicate the change of variable $y = \alpha\,\psi(x)\,t$. For $t > 0$ the fields $U_\alpha^{(n)}$ and $V_\alpha^{(n)}$ satisfy the second order parabolic systems

$$\partial_t U_\alpha^{(n)} - \Delta U_\alpha^{(n)} = L_1^{1,1} U_\alpha^{(n)} + L_1^{1,2} V_\alpha^{(n)} + B_1^1 \big(U_\alpha^{(n-1)}, V_\alpha^{(n-1)}\big)$$

$$+ \widetilde{L}^{1,1}\big(P_\alpha^{(n-1)}\big) + \widetilde{L}^{1,2}\big(\Pi_\alpha^{(n-1)}\big) + F_\alpha + \Phi_1,$$

and

$$\partial_t V_\alpha^{(n)} - \Delta V_\alpha^{(n)} = L_1^{2,1} U_\alpha^{(n)} + L_1^{2,2} V_\alpha^{(n)} + B_1^2 (U_\alpha^{(n-1)}, V_\alpha^{(n-1)})$$

$$+ \widetilde{L}^{2,1} (P_\alpha^{(n-1)}) + \widetilde{L}^{2,2} (\Pi_\alpha^{(n-1)}) + G_\alpha,$$

where the coefficients on the right hand sides are matrices of differential operators which are precisely constructed in Bradshaw, Grujić, & Kukavica (2015). Similarly, at each time $t \in (0, T_0)$, the pressure terms $P_\alpha^{(n)}$ and $\Pi_\alpha^{(n)}$ satisfy the elliptic systems

$$-\Delta P_\alpha^{(n)} = L_2^{1,1} (P_\alpha^{(n)}) + L_2^{1,2} (\Pi_\alpha^{(n)}) + B_2^1 (U_\alpha^{(n-1)}, V_\alpha^{(n-1)}) + \Phi_2,$$

and

$$-\Delta \Pi_\alpha^{(n)} = L_2^{2,1} (P_\alpha^{(n)}) + L_2^{2,2} (\Pi_\alpha^{(n)}) + B_2^2 (U_\alpha^{(n-1)}, V_\alpha^{(n-1)}).$$

An inductive argument leads to uniform bounds on the $L^q(2B_*)$ norms of $U_\alpha^{(n)}$ and $V_\alpha^{(n)}$ provided T and α are suitably controlled. Let $\widetilde{U}^{(n)} = U^{(n)} - \phi u$. It is possible to show that if $T < T_1$ and $|\alpha| t \le C_0^{-1}$ for a universal constant C_0 and, for some $n \in \mathbb{N}$, we have

$$\sup_{0<t<T} \|\widetilde{U}_\alpha^{(n)}(t)\|_{L^q(2B_*)} + \sup_{0<t<T} \|V_\alpha^{(n)}(t)\|_{L^q(2B_*)} \le 4 M_q, \qquad (2.1)$$

and

$$T^{(r-2)/(2r)} \left[\left(\int_0^T \|\nabla \widetilde{U}_\alpha^{(n)}(t)\|_{L^q(2B_*)}^r \, dt \right)^{\frac{1}{r}} \right. \qquad (2.2)$$

$$\left. + \left(\int_0^T \|\nabla V_\alpha^{(n)}(t)\|_{L^q(2B_*)}^r \, dt \right)^{\frac{1}{r}} \right] \le 4 M_q, \quad (2.3)$$

then, also,

$$\sup_{0<t<T} \|\widetilde{U}_\alpha^{(n+1)}(t)\|_{L^q(2B_*)} + \sup_{0<t<T} \|V_\alpha^{(n+1)}(t)\|_{L^q(2B_*)} \le 4 M_q, \quad (2.4)$$

and

$$T^{(r-2)/(2r)} \left[\left(\int_0^T \|\nabla \widetilde{U}_\alpha^{(n+1)}(t)\|_{L^q(2B_*)}^r \, dt \right)^{\frac{1}{r}} \right. \qquad (2.5)$$

$$\left. + \left(\int_0^T \|\nabla V_\alpha^{(n+1)}(t)\|_{L^q(2B_*)}^r \, dt \right)^{\frac{1}{r}} \right] \le 4 M_q. \quad (2.6)$$

For $|\alpha| \le (C_0 T_1)^{-1}$ and $t < T$ it follows that

$$\sup_{0<t<T} \|U_\alpha^{(n+1)}(t)\|_{L^q(2B_*)} + \sup_{0<t<T} \|V_\alpha^{(n+1)}(t)\|_{L^q(2B_*)} \le 5 M_q,$$

from which we obtain a uniform bound on the complex integrals

$$\int_{\Omega_*(t)} |U^{(n)} + iV^{(n)}|^q \, dx \, dy.$$

Consequently, $\{(U^{(n)} + iV^{(n)})|_{\Omega_*(t)}\}$ is a normal family. Applying a convergence argument contained in Grujić & Kukavica (1998) concludes the proof of Theorem 2.3.1. $\square$

2.4 A conditional regularity criteria in weak-L^3

We will devote this section to establishing a regularity criteria in the critical space $L^\infty(0, T; L^3_w(\mathbb{R}^3))$. The proof uses properties of the analyticity radius of a mild solution with initial data in L^∞. These are contained in Guberović (2010); we recall the main result for convenience.

Theorem 2.4.1 (Guberović (2010)). *Let u_0 be in $L^\infty(\mathbb{R}^3)$, and let u be the unique mild solution evolving from u_0. There exists an absolute constant $c_0 > 1$ so that, for $T = 1/c_0^2\|u_0\|_\infty^2$ and $t \in (0, T]$, the solution $u(t)$ has an analytic extension $U(t) + iV(t)$ to the region*

$$D_t = \left\{ x + iy \in \mathbb{C}^3 : |y| \leq \frac{1}{c_0}\sqrt{t} \right\}.$$

In addition,

$$\|U(t)\|_{L^\infty(D_t)}, \|V(t)\|_{L^\infty(D_t)} \leq c_0\|u_0\|_\infty,$$

for all t in $(0, T]$.

In Grujić (2013), a geometric measure type regularity criteria is developed using a notion of sparseness which we presently define.

Definition 2.4.2. Let x_0 be a point in $\mathbb{R}^3$, $r > 0$, S an open subset of $\mathbb{R}^3$ and δ in $(0, 1)$. The set S is *linearly δ-sparse around x_0 at scale r* if there exists a unit vector d in S^2 such that

$$\frac{|S \cap (x_0 - rd, x_0 + rd)|}{2r} \leq \delta.$$

Denote the region of intense fluid activity at time t by $S(t)$ and the lower bound on the uniform analyticity radius by $\rho(t)$. The conditional regularity criteria in Grujić (2013) requires that, at finitely many times close to a possible blow-up time, $S(t)$ is sparse in the sense of Definition 2.4.2 at scales smaller than $\rho(t)$. The main mathematical device behind this result is a pointwise interpolative estimate which we include as a lemma.

Lemma 2.4.3. *Let $f\colon \mathbb{R}^3 \to \mathbb{R}^3$ be real analytic at x_0. Denote its analyticity radius by ρ_0. Let B_r denote the complex ball centered at x_0 of radius r and denote by $F+iG$ the analytic extension of f to B_{ρ_0}. Suppose $x_0 \in S = \{x \in \mathbb{R}^3 : |f(x)| \geq M\}$. If S is linearly δ-sparse around x_0 for the length scale $r \leq \rho_0$ in the direction d for some $\delta \in (0,1)$, then*

$$|f(x_0)| \leq m^{1-h} M^h,$$

where

$$h = h(\delta) = \frac{2}{\pi} \arcsin \frac{1-\delta^2}{1+\delta^2},$$

and $m = \|F\|_{L^\infty(B_r)} + \|G\|_{L^\infty(B_r)}$.

This lemma can be applied in an amorphous scenario, i.e. one where sparseness is not driven by anisotropic geometric features, to obtain a new and simple proof of the regularity at time T of solutions with sufficiently small $L^\infty(0, T; L_w^3(\mathbb{R}^3))$-norm.

Theorem 2.4.4. *Let u be a solution to 3D NSE which is regular on $(0, T)$. There exists an absolute constant ϵ_0 so that if*

$$K = \sup_{0<t<T} \|u(t)\|_{L_w^3(\mathbb{R}^3)} < \epsilon_0,$$

then u is regular at T.

Proof. For $t_1, t_2 \in (0, T)$ let

$$S(t_1, t_2) = \left\{ (x, t_2) : |u(x, t_2)| > \frac{1}{c_0^2} \|u(t_1)\|_\infty \right\},$$

where c_0 is the constant appearing in Theorem 2.4.1. We estimate $|S(t_1, t_2)|$ using the L_w^3 norm of $u(t_2)$ we obtain

$$|S(t_1, t_2)|^{1/3} \leq \frac{c_0^2 K}{\|u(t_1)\|_\infty}.$$

Let

$$t_2 = t_1 + \frac{1}{2c_0^2 \|u(t_1)\|_\infty^2}.$$

Denote the analyticity radius of $u(t_2)$ by $\rho(t_2)$. By Theorem 2.4.1 we know $\rho(t_2) \geq (c_0^2 \|u(t_1)\|_\infty)^{-1}$. Choosing $\epsilon_0 < (3\, c_0^4)^{-1}$ ensures that

$$|S(t_1, t_2)|^{1/3} \leq \frac{\rho(t_2)}{3},$$

and thus $S(t_1, t_2)$ is linearly δ-sparse at scale $\rho(t_2)$ in some direction at every $x \in S(t_1, t_2)$ (we are taking $\delta = 1/3$). Applying Lemma 2.4.3 at each $x \in S(t_1, t_2)$ yields the bound

$$|u(x, t_2)| \leq \|u(t_1)\|_{L^\infty}.$$

This bound clearly extends to the complement of $S(t_1, t_2)$ and we conclude that

$$\|u(x, t_2)\|_{L^\infty} \leq \|u(t_1)\|_{L^\infty}.$$

Consider the times $t_1, \ldots, t_k$ which occur prior to T and satisfy

$$t_n = t_{n-1} + \frac{1}{2c_0^2 \|u(t_1)\|_\infty^2}.$$

Iteratively applying this procedure to the sets $S(t_1, t_n)$ ensures that

$$\|u(t_n)\|_{L^\infty} \leq \|u(t_1)\|_{L^\infty}.$$

Using Theorem 2.4.1, re-solving at time t_k provides a smooth extension of u beyond T indicating u is regular at T. $\square$

Acknowledgments

The collaboration between Z.B. and I.K. were accommodated through the American Mathematical Society's Mathematical Research Communities – 2013 program on "Regularity Problems for Nonlinear Partial Differential Equations Modeling Fluids and Complex Fluids." Z.B. acknowledges the support of the *Virginia Space Grant Consortium* via the Graduate Research Fellowship; Z.G. acknowledges the support of the *Research Council of Norway* via the grant 213474/F20 and the *National Science Foundation* via the grant DMS 1212023; I.K. was supported in part by the *National Science Foundation* grant DMS 1311943.

References

Bae, H., Biswas, A., & Tadmor, E. (2012) Analyticity and decay estimates of the Navier–Stokes equations in critical Besov spaces. *Arch. Ration. Mech. Anal.* **205**, no. 3, 963–991.

Bae, H., Biswas, A., & Tadmor, E. (to appear) Analyticity of the subcritical and critical quasi-geostrophic equations in Besov Spaces. (arXiv:1310.1624)

Biswas, A. (2005) Local existence and Gevrey regularity of 3-D Navier–Stokes equations with l^p initial data. *J. Differential Equations* **215**, no. 2, 429–447.

Biswas, A. & Foias, C. (2014) On the maximal space analyticity radius for the 3D Navier–Stokes equations and energy cascades. *Ann. Mat. Pura Appl.* **193**, no. 3, 739–777.

Biswas, A., Jolly, M., Martinez, V., & Titi, E.S. (2014) Dissipation length scale estimates for turbulent flows: a Wiener algebra approach. *J. Nonlinear Sci.* **24**, no. 3, 441–471.

Biswas, A. & Swanson, D. Gevrey regularity of solutions to the 3-D Navier–Stokes equations with weighted l_p initial data. *Indiana Univ. Math. J.* **56**, no. 3, 1157–1188.

Bradshaw, Z. & Grujić, Z. (2014) Blow-up scenarios for 3D NSE exhibiting sub-criticality with respect to the scaling of one-dimensional local sparseness. *J. Math. Fluid Mech.* **16**, 321–334.

Bradshaw, Z. & Grujić, Z. A spatially localized $L \log L$ estimate on the vorticity in the 3D NSE. *Indiana Univ. Math. J.* **64**, 433–440.

Bradshaw, Z., Grujić, Z., & Kukavica, I. Local analyticity radii of solutions to the 3D Navier–Stokes equations with locally analytic forcing. *J. Differential Equations* **259**, 3955–3975.

Escauriaza, L., Seregin, G., & Šverák, V. (2003) $L_{3,\infty}$-solutions of Navier–Stokes equations and backward uniqueness. *Uspekhi Mat. Nauk* **58**, no. 2(350), 3–44.

Ferrari, A. & Titi, E. (1998) Gevrey regularity for nonlinear analytic parabolic equations. *Comm. Partial Differential Equations* **23**, no. 1-2, 1–16.

Foias, C. (1995) What do the Navier–Stokes equations tell us about turbulence? in *Harmonic analysis and nonlinear differential equations* (Riverside, CA). Contemp. Math., 208, Amer. Math. Soc., Providence, RI, 151–180.

Foias, C. & Temam, R. (1989) Gevrey class regularity for the solutions of the Navier–Stokes equations. *J. Funct. Anal.* **87**, no. 2, 359–369.

Frisch, U. (1995) *Turbulence, the legacy of A. N. Kolmogorov.* Cambridge University Press, Cambridge, UK.

Gevrey, M. (1918) Sur la nature analytique des solutions des équations aux dérivées partielles. *Annales Scientifiques de l'Ecole Normale Supérieure* **35**, 129–190.

Giga, Y. (1983) Time and spatial analyticity of solutions of the Navier–Stokes equations. *Comm. Partial Differential Equations* **8**, no. 8, 929–948.

Grujić, Z. (2013) A geometric measure-type regularity criterion for solutions to the 3D Navier–Stokes equations. *Nonlinearity* **26**, no. 1, 289–296.

Grujić, Z. & Kukavica, I. (1998) Space analyticity for the Navier–Stokes and related equations with initial data in L^p. *J. Funct. Anal.* **152**, no. 2, 447–466.

Grujić, Z. & Kukavica, I. (1999) Space analyticity for the nonlinear heat equation in a bounded domain. *J. Differential Equations* **154**, no. 1, 42–54.

Guberović, R. (2010) Smoothness of Koch–Tataru solutions to the Navier–Stokes equations revisited. *Discrete Contin. Dyn. Syst.* **27**, no. 1, 231–236.

Henshaw, W., Kreiss, H.O., & Reyna, L. (1990) Smallest scale estimates for the Navier–Stokes equations for incompressible fluids. *Arch. Rational Mech. Anal.* **112**, no. 1, 21–44.

Henshaw, W., Kreiss, H.O., & Reyna, L. (1995) Estimates of the local minimum scale for the incompressible Navier–Stokes equations. *Numer. Funct. Anal. Optim.* **16**, no. 3-4, 315–344.

Ignatova, M., Kukavica, I., & Ziane, M. (2012) Local existence of solutions to the free boundary value problem for the primitive equations of the ocean. *J. Math. Phys.* **53**, 103101.

Kahane, C. (1969) On the spatial analyticity of solutions of the Navier–Stokes equations. *Arch. Rational Mech. Anal.* **33**, 386–405.

Kato, T. (1984) Strong L^p-solutions of the Navier–Stokes equation in $\mathbb{R}^m$, with applications to weak solutions. *Math. Z.* **187**, no. 4, 471–480.

Kim, H. & Kozono, H. (2006) A removable isolated singularity theorem for the stationary Navier–Stokes equations. *J. Differential Equations* **220**, no. 1, 68–84.

Kim, H. & Kozono, H. (2004) Interior regularity criteria in weak spaces for the Navier–Stokes equations. *Manuscripta Math.* **115**, no. 1, 85–100.

Komatsu, G. (1979) Analyticity up to the boundary of solutions of nonlinear parabolic equations. *Comm. Pure Appl. Math.* **32**, no. 5, 669–720.

Kukavica, I. (1999) On the dissipative scale for the Navier–Stokes equation. *Indiana Univ. Math. J.* **48**, no. 3, 1057–1081.

Kukavica, I. & Vicol, V. (2009) On the radius of analyticity of solutions to the three-dimensional Euler equations. *Proc. Amer. Math. Soc.* **137**, no. 2, 669–677.

Kukavica, I. & Vicol, V. (2011b) The domain of analyticity of solutions to the three-dimensional Euler equations in a half space. *Discrete Contin. Dyn. Syst.* **29**, no. 1, 285–303.

Lemarié-Rieusset, P.G. (2000) Une remarque sur l'analyticité des solutions milds des équations de Navier–Stokes dans $\mathbb{R}^3$. *C. R. Acad. Sci. Paris Sér. I Math.* **330**, no. 3, 183–186.

Lemarié-Rieusset, P.G. (2002) *Recent developments in the Navier–Stokes problem*. Chapman and Hall/CRC Research Notes in Mathematics **431**. Chapman and Hall/CRC, Boca Raton, FL.

Lemarié-Rieusset, P.G. (2004) Nouvelles remarques sur l'analyticité des solutions milds des équations de Navier–Stokes dans $\mathbb{R}^3$. *C. R. Math. Acad. Sci. Paris* **338**, no. 6, 443–446.

Levermore, C.D. & Oliver, M. (1997) Analyticity of solutions for a generalized Euler equation. *J. Differential Equations* **133**, no. 2, 321–339.

Luo, Y. & Tsai, T.P. (2013) Regularity criteria in weak L^3 for 3D incompressible Navier–Stokes equations. arXiv:1310.8307.

Masuda, K. (1967) On the analyticity and the unique continuation theorem for solutions of the Navier–Stokes equation. *Proc. Japan Acad.* **43**, 827–832.

Monin, A. & Yaglom, A. (1971) *Statistical Fluid Mechanics: Mechanics of Turbulence*, Volume II Cambridge, Mass, MIT Press.

Morrey, C. (1958) On the analyticity of the solutions of analytic non-linear elliptic systems of partial differential equations. *Amer. J. Math.* **80**, 198–237.

Nakagawa, K. (1981) On the analyticity of the solutions of the Navier–Stokes equations. *Tohoku Math. J.* **33**, no. 2, 177–192.

Oliver, M. & Titi, E.S. (2001) On the domain of spatial analyticity for solutions of second order nonlinear analytic parabolic and elliptic differential equations. *J. Differential Equations* **174**, 55–74.

Oliver, M. & Titi, E.S. (2000) Remark on the rate of decay of higher order derivatives for solutions to the Navier–Stokes equations in $\mathbb{R}^n$. *J. Funct. Anal.* **172**, no. 1, 1–18.

Paciu, M. & Vicol, V. (2011) Analyticity and Gevrey-class regularity for the second-grade fluid equations. *J. Math. Fluid Mech.* **13**, no. 4, 533–555.

Serrin, J. (1962) On the interior regularity of weak solutions of the Navier–Stokes equations. *Arch. Rational Mech. Anal.* **9**, 187–195.

3

On the motion of a pendulum with a cavity entirely filled with a viscous liquid

Giovanni P. Galdi & Giusy Mazzone

*Department of Mechanical Engineering and Materials Science,
University of Pittsburgh, PA 15261.*
`galdi@pitt.edu`, `gim20@pitt.edu`

3.1 Introduction

As is well known, a (physical) pendulum is a system characterized by a heavy rigid body, $\mathscr{B}$, constrained to rotate around a horizontal axis, a, so that its centre of mass C, satisfies the following properties: (i) the distance, ℓ, between C and its orthogonal projection O on a (the *point of suspension*), does not depend on time, and (ii) C always moves in a plane orthogonal to a. It is a classical result that in absence of friction (an assumption that we shall keep throughout the paper) the generic motion of $\mathscr{B}$ is a "nonlinear oscillation" and, in particular, motions of "small amplitude" around the lowest position of C are undamped oscillations with frequency $\sqrt{m\,g\,\ell/I}$, where g is the acceleration of gravity and m and I represent the mass of $\mathscr{B}$ and its moment of inertia around a, respectively; see Chapter III in Pook (2011), for example.

Suppose now that there is a hollow cavity in $\mathscr{B}$ filled completely with a viscous liquid. Experimental evidence then shows that the liquid will have a stabilizing effect on the motion of the pendulum, by reducing the amplitude of oscillations. A most remarkable application of this property occurs in space engineering, where tube dampers filled with a viscous liquid are used to suppress oscillations in spacecraft and artificial satellites; see, for example, Alfriend & Spencer (1983), Bhuta & Koval (1966), Boyevkin et al. (1976), Sarychev (1978) and the literature there cited.

Despite the numerous contributions aimed at furnishing a mathematical analysis of this very interesting phenomenon (including Chapter 1 §6 in Chernousko (1972), Chapter 7.5 in Kopachevsky & Krein (2000),

Published in *Recent Progress in the Theory of the Euler and Navier-Stokes Equations*, edited by James C. Robinson, José L. Rodrigo, Witold Sadowski, & Alejandro Vidal-López. ©Cambridge University Press 2016.

Krasnoshchekov (1963), p. 182ff in Moiseyev & Rumyantsev (1968), Pivovarov & Chernousko (1990), and Smirnova (1975)), those results established so far cannot be considered fully satisfactory. In fact, they are not of an exact nature, due either to the approximate models used or else to an approximate mathematical treatment, such as, for instance, assuming large viscosity limit, special shapes of the cavity, and/or employing suitable linearizations of the original equations.

The objective of this paper is to provide a rigorous analysis of the motion of the coupled system, $\mathscr{S}$, constituted by a pendulum $\mathscr{B}$ with an interior cavity, $\mathscr{C}$, completely filled with a viscous (Navier–Stokes) liquid, $\mathscr{L}$. In particular, we shall show that, provided $\mathscr{C}$ is sufficiently regular, all motions of $\mathscr{S}$ described within a very general class of solutions to the relevant equations (*weak solutions*), must tend to a rest state for large times, no matter the shape of $\mathscr{C}$, the physical characteristics of $\mathscr{B}$ and $\mathscr{L}$, and the initial conditions imparted to $\mathscr{S}$. We show that, as expected, the rest state is realized by only two equilibrium configurations of $\mathscr{S}$, namely, those where the velocity field of $\mathscr{L}$ is zero, and the centre of mass G of $\mathscr{S}$ is in its lowest, G_l, or highest, G_h, position; see Theorem 3.4.4. We then further prove that for a broad set of initial data, the final state must be the one with $G \equiv G_l$. This set includes the case when the system $\mathscr{S}$ is released from rest; see Theorem 3.5.1. In physical terms, the latter translates into the following interesting property, namely, that a pendulum with a *cavity filled with a viscous liquid* that is initially at rest *eventually reaches the equilibrium configuration where the centre of mass is at its lowest point, the same as happens to a classical pendulum immersed in a viscous liquid.* However, it must be also observed that the *global* dynamics can be quite different in the two cases. In fact, while in the latter the amplitude of oscillations may gradually decrease from the outset till it reduces to zero, in the former, in analogy to similar problems of solids with liquid-filled cavity Leung & Kuang (2007), Galdi, Mazzone, & Zunino (2013), the damping of the oscillations may take place only after an interval of time $[0, T]$, say, where, possibly, a motion of "chaotic" nature occurs, with T depending on the magnitude of the viscosity ν ($T \to \infty$ as $\nu \to 0$).

The method we use to show the above results is the one we announced in Galdi et al. (2013) and developed in Galdi, Mazzone, & Zunino (2014), which employs tools from the classical theory of dynamical systems. However, the adaptation of these tools to our problem is not trivial in that we deal with weak solutions (à la Leray–Hopf) where the uniqueness property is not guaranteed. Nevertheless, we are able to show that every such solution *eventually* becomes "strong" and this is enough to provide a complete characterization of the associated Ω-limit set; see Proposition 3.3.6, Proposition 3.4.2, and Proposition 3.4.3.

The plan of the paper is the following. After recalling some classical notation in Section 2, in Section 3 we give a mathematical formulation of the problem and introduce the class of weak solutions. Moreover, we study the property of these solutions and show, in particular, that they become regular for sufficiently large times (Proposition 3.3.6). In the following Section 4, we characterize the Ω-limit set of weak solutions corresponding to initial data having finite kinetic energy, and show that they must tend, for large times, to one of the two possible equilibrium configurations, namely, when the liquid and the body are at rest and the centre of mass of the system lies at its lowest/highest position (Proposition 3.4.3 and Theorem 3.4.4). In Section 5, we prove that only the former equilibrium can be attained from a large class of initial data (Theorem 3.5.1), and is stable (Theorem 3.5.2). Moreover, we show that the other equilibrium configuration is always unstable (Theorem 3.5.2). Finally, in Section 7, we provide some concluding remarks, and point out certain interesting open questions.

3.2 Notation

The symbol $\mathbb{N}$ denotes the set of positive integers. By $\mathbb{R}$ we denote the set of real numbers, and by $\mathbb{R}^3$ the Euclidean three-dimensional space. Vectors in $\mathbb{R}^3$ will be indicated by boldfaced letters, and the canonical basis, B, in $\mathbb{R}^3$ by $\{e_1, e_2, e_3\}$. Components of a vector v in B are indicated by (v_1, v_2, v_3), whereas $|v|$ represents the magnitude of v.

Let A be a domain of $\mathbb{R}^3$. We denote by $L^2(A)$, and $W^{k,2}(A)$, $W_0^{k,2}(A)$, $k \in \mathbb{N}$, the usual Lebesgue and Sobolev spaces, with norms $\|\cdot\|_2$ and $\|\cdot\|_{k,2}$, respectively.[1]

If $\{\boldsymbol{G}, \boldsymbol{H}\}$, $\{\boldsymbol{g}, \boldsymbol{h}\}$, $\{g, h\}$ are pairs of second-order tensor, vector and scalar fields on A, respectively, we set

$$(\boldsymbol{G}, \boldsymbol{H})_A = \int_A G_{ij} H_{ij}, \quad (\boldsymbol{g}, \boldsymbol{h})_A = \int_A g_i h_i, \quad (g, h)_A = \int_A g\,h,$$

where, typically, we shall omit the subscript A. (We use the Einstein summation convention and sum over repeated indices.)

---

[1]Unless confusion arises, we shall use the same symbol for spaces of scalar, vector and tensor functions. Moreover, in integrals we shall typically omit the infinitesimal element of integration.

By $H(A)$ we denote the completion of divergence-free smooth functions with a compact support in A. If A is bounded and Lipschitz with outward unit normal n, we have Galdi (2011)

$$H(A) = \{u \in L^2(A) : \operatorname{div} u = 0 \text{ and } u \cdot n|_{\partial A} = 0\},$$

where $\operatorname{div} u$ and $u \cdot n|_{\partial A}$ are understood in the sense of distributions.

If X is a Banach space with norm $\| \cdot \|_X$, and $I \subset \mathbb{R}$ an interval, we denote by $L^q(I; X)$ [respectively $W^{k,q}(I; X)$, $k \in \mathbb{N}$], the space of functions $f : I \mapsto X$ such that $\left(\int_I \|f(t)\|_X^q dt\right)^{1/q} < \infty$ [respectively $\sum_{l=0}^{k} \left(\int_I \|\partial_t^l f(t)\|_X^q dt\right)^{1/q} < \infty$]. Likewise, we write $f \in C^k(I; X)$ if f is k-times differentiable with values in X and $\max_{t \in I} \|\partial_t^l f(t)\|_X < \infty$, $l = 0, 1, \ldots, k$. Moreover, $f \in C_w(I; X)$ means that the map f is weakly continuous from I into X, i.e. $t \in I \to \ell(f(t)) \in \mathbb{R}$ for all bounded linear functionals ℓ defined on X. In denoting all the above spaces, we shall omit the symbol X if $X = \mathbb{R}$.

3.3 Governing equations. Weak solutions and preliminary results

Consider the coupled system, $\mathscr{S}$, constituted by a rigid body, $\mathscr{B}$, with an interior cavity, $\mathscr{C}$, entirely filled with a Navier–Stokes liquid, $\mathscr{L}$. Suppose that $\mathscr{B}$ is constrained to move (without friction) around a horizontal axis a in such a way that the centre of mass G of $\mathscr{S}$ satisfies the following properties during all possible motions of $\mathscr{S}$:

(i) the distance from G to its orthogonal projection, O, on a is constant, and

(ii) G belongs to a fixed vertical plane orthogonal to a.

Let $\mathscr{F} \equiv \{O, e_1, e_2, e_3\}$ be a frame attached to $\mathscr{B}$, with the origin at O, $e_1 \equiv \overrightarrow{OG} / |\overrightarrow{OG}|$, and e_3 directed along a. Then the motion of $\mathscr{S}$ in $\mathscr{F}$ is governed by the following set of equations in $\mathscr{C} \times \mathbb{R}_+$ (see Moiseyev & Rumyantsev, 1968)

$$\left. \begin{aligned} \rho\big(v_t + \big(\dot{a} + \beta^2 \gamma_2\big) e_3 \times x + v \cdot \nabla v + 2\omega\, e_3 \times v\big) &= \mu \Delta v - \nabla p, \\ \nabla \cdot v &= 0, \end{aligned} \right\}$$

$$v(x, t)|_{\partial \mathscr{C}} = 0,$$

$$\dot{\omega} - \dot{a} = \beta^2 \gamma_2, \quad \dot{\gamma} + \omega\, e_3 \times \gamma = 0.$$

$$(3.1)$$

Here, $\boldsymbol{v}$ and p are the *relative* velocity and the (modified) pressure fields of $\mathscr{L}$, respectively, while ρ and μ are its density and shear viscosity coefficient. Also, $\omega\,\boldsymbol{e}_3$ is the angular velocity of $\mathscr{B}$ and $\boldsymbol{\gamma} = (\gamma_1, \gamma_2, 0) = (\cos\varphi, -\sin\varphi, 0)$ where φ is the angle between $\boldsymbol{e}_1$ and the gravity vector $\boldsymbol{g}$. Furthermore,

$$a := -\frac{\rho}{C}\,\boldsymbol{e}_3 \cdot \int_{\mathscr{C}} \boldsymbol{x} \times \boldsymbol{v}, \tag{3.2}$$

where C is the moment of inertia of $\mathscr{S}$ with respect to a, and

$$\beta^2 = M\,g\,|\,\overrightarrow{OG}\,|/C,$$

with M mass of $\mathscr{S}$.

Remark 3.3.1. In problem (3.1), the unknowns are $\boldsymbol{v}, p, \omega$, and $\boldsymbol{\gamma}$. The vector function $\boldsymbol{\gamma}$ can be interpreted as the direction of gravity, which thus becomes time dependent when observed from the body-fixed frame $\mathscr{F}$. In this regard, we note that if we write $\omega = \dot{\varphi}$, it follows that the equation $(3.1)_5$ for $\boldsymbol{\gamma}$ is automatically satisfied, and problem (3.1) can be rewritten in a mixed Eulerian-Lagrangian form where the unknowns now become $\boldsymbol{v}, p$, and φ. However, we prefer to use the formulation presented above, since it is particularly suitable from the viewpoint of our approach.

An important consequence of (3.1) is the equation of *energy balance* that we derive next. By dot-multiplying $(3.1)_1$ by $\boldsymbol{v}$, integrating by parts over $\mathscr{C}$ and taking into account $(3.1)_{2,3}$ and (3.2) we show that

$$\tfrac{1}{2}\frac{\mathrm{d}}{\mathrm{d}t}\left(\rho\,\|\boldsymbol{v}\|_2^2 - C\,a^2\right) + \mu\,\|\nabla\boldsymbol{v}\|_2^2 = C\,\beta^2\gamma_2 a. \tag{3.3}$$

On the other hand, from the last two equations in (3.1) we infer that

$$\tfrac{1}{2}\frac{\mathrm{d}}{\mathrm{d}t}C\,(\omega - a)^2 = -C\,\beta^2\gamma_2 a + C\,\beta^2\gamma_2\omega = -C\,\beta^2\gamma_2 a + C\beta^2\frac{d\gamma_1}{dt}. \tag{3.4}$$

Summing side-by-side (3.3) and (3.4) we then deduce the desired energy balance equation

$$\tfrac{1}{2}\frac{\mathrm{d}}{\mathrm{d}t}\left[\rho\,\|\boldsymbol{v}\|_2^2 - C\,a^2 + C\,(\omega - a)^2 - 2C\beta^2\gamma_1\right] + \mu\,\|\nabla\boldsymbol{v}\|_2^2 = 0. \tag{3.5}$$

In this equation, the quantity

$$\mathcal{E} := \tfrac{1}{2}\left[\rho\,\|\boldsymbol{v}\|_2^2 - C\,a^2 + C\,(\omega - a)^2\right] \tag{3.6}$$

represents the total kinetic energy of $\mathscr{S}$, while

$$\mathcal{U} := -C\beta^2\gamma_1 \tag{3.7}$$

is its potential energy. One can show that there is a positive constant $c_0 \leq 1$, such that

$$c_0 \left(\rho \|\boldsymbol{v}\|_2^2 + C\left(\omega - a\right)^2\right) \leq 2\mathcal{E} \leq \left(\rho \|\boldsymbol{v}\|_2^2 + C\left(\omega - a\right)^2\right). \qquad (3.8)$$

In fact, the bound from above is trivial, while the one from below follows from the following inequality, a special case of that proved in §7.2.2 of Kopachevsky & Krein (2000)

$$c_0 \|\boldsymbol{w}\|_2^2 \leq \left[\|\boldsymbol{w}\|_2^2 - \frac{\rho}{C}\left(\boldsymbol{e}_3 \cdot \int_{\mathscr{C}} \boldsymbol{x} \times \boldsymbol{w}\right)^2\right], \quad \boldsymbol{w} \in H(\mathscr{C}). \qquad (3.9)$$

One of our main goals is to study the asymptotic behaviour in time of the coupled system $\mathscr{S}$ in the very general class constituted by weak solutions (à la Leray-Hopf) to (3.1). To this end, we give the following definition. As customary, S^1 denotes the unit sphere in $\mathbb{R}^2$.

Definition 3.3.2. The triple $(\boldsymbol{v}, \omega, \boldsymbol{\gamma})$ is a *weak solution* to (3.1) if it meets the following requirements:

(a) $\boldsymbol{v} \in C_w([0, \infty); H(\mathscr{C})) \cap L^\infty(0, \infty; H(\mathscr{C})) \cap L^2(0, \infty; W_0^{1,2}(\mathscr{C}))$;

(b) $\omega \in C^0([0, \infty)) \cap L^\infty(0, \infty)$, $\boldsymbol{\gamma} \in C^1([0, \infty); \mathsf{S}^1)$;

(c) Strong Energy Inequality:

$$\mathcal{E}(t) + \mathcal{U}(t) + \mu \int_s^t \|\nabla \boldsymbol{v}(\tau)\|_2^2 \, d\tau \leq \mathcal{E}(s) + \mathcal{U}(s) \qquad (3.10)$$

for all $t \geq s$ and a.e. $s \geq 0$ including $s = 0$;

(d) $(\boldsymbol{v}, \omega, \boldsymbol{\gamma})$ satisfies $(3.1)_{1,2,4,5}$ in the sense of distributions and $(3.1)_3$ in the trace sense.

The next result ensures that, provided the initial data have finite energy, there exists at least one corresponding weak solution.

Proposition 3.3.3. *Let $\mathscr{C}$ be a bounded domain in $\mathbb{R}^3$. Then, for any given*

$$\boldsymbol{v}_0 \in H(\mathscr{C}), \quad \omega_0 \in \mathbb{R}, \quad \boldsymbol{\gamma}_0 \in \mathsf{S}^1,$$

there exists at least one weak solution to (3.1) such that

$$\lim_{t \to 0^+} \|\boldsymbol{v}(t) - \boldsymbol{v}_0\|_2 = \lim_{t \to 0^+} |\omega(t) - \omega_0| = \lim_{t \to 0^+} |\boldsymbol{\gamma}(t) - \boldsymbol{\gamma}_0| = 0.$$

This existence result can be accomplished by a classical procedure consisting of combining the standard Galerkin method with suitable energy estimates. In fact, it is enough to introduce the field $\boldsymbol{u} := \boldsymbol{v} + \omega \boldsymbol{e}_3 \times \boldsymbol{x}$ (*absolute* velocity), and proceed as in Mazzone (2012). Since, with the exception of minor changes, the existence proof is entirely analogous to (and, to some extent, simpler than) that given in Chapter 3 of Mazzone (2012) (see also Theorem 5.6 in Silvestre & Takahashi, 2012) it will be omitted.

As for uniqueness, in analogy with the ordinary Navier–Stokes initial-value problem, we do not know whether a weak solution is unique. However, we have the following result for whose proof we refer again to Mazzone (2012, §3.4).

Proposition 3.3.4. *Let* $(\boldsymbol{v}, \omega, \boldsymbol{\gamma})$ *and* $(\boldsymbol{v}^*, \omega^*, \boldsymbol{\gamma}^*)$ *be two weak solutions corresponding to initial data* $(\boldsymbol{v}_0, \omega_0, \boldsymbol{\gamma}_0)$ *and* $(\boldsymbol{v}_0^*, \omega_0^*, \boldsymbol{\gamma}_0^*)$, *respectively. Suppose there is* $T > 0$ *such that*

$$\boldsymbol{v}^* \in L^q(0, T; L^r(\mathscr{C})), \quad \frac{2}{q} + \frac{3}{r} = 1, \quad \text{some } r > 3. \qquad (3.11)$$

Then for all $t \in [0, T]$

$$\|\boldsymbol{v}(t) - \boldsymbol{v}^*(t)\|_2 + |\omega(t) - \omega^*(t)| + |\boldsymbol{\gamma}(t) - \boldsymbol{\gamma}^*(t)|$$
$$\leq c \left(\|\boldsymbol{v}_0 - \boldsymbol{v}_0^*\|_2 + |\omega_0 - \omega_0^*| + |\boldsymbol{\gamma}_0 - \boldsymbol{\gamma}_0^*| \right),$$

where c *depends on* $\operatorname*{ess\,sup}_{t \in [0,T]} \|\boldsymbol{v}(t)\|_2$, $\operatorname*{ess\,sup}_{t \in [0,T]} \|\boldsymbol{v}^*(t)\|_2$, $\|\boldsymbol{v}^*\|_{L^q(0,T;L^r(\mathscr{C}))}$, *and* $\max_{t \in [0,T]} |\omega^*(t)|$. *Thus, in particular, if* $\boldsymbol{v}_0 = \boldsymbol{v}_0^*$, $\omega_0 = \omega_0^*$, *and* $\boldsymbol{\gamma}_0 = \boldsymbol{\gamma}_0^*$, *it follows that* $(\boldsymbol{v}, \omega, \boldsymbol{\gamma}) \equiv (\boldsymbol{v}^*, \omega^*, \boldsymbol{\gamma}^*)$ *a.e. in* $\mathscr{C} \times [0, T]$.

We conclude this section by showing an important property that implies, in particular, that every weak solution becomes strong (in a suitable sense) for all sufficiently large times. To this end, we shall use the following lemma whose proof can be found in the Appendix of Galdi et al. (2014).

Lemma 3.3.5. *Let* $y : [t_0, t_1) \to [0, \infty)$, $t_1 > t_0 \geq 0$, *be an absolutely continuous function satisfying, for some* $a, b, c > 0$, *and* $\alpha > 1$,

(i) $y' \leq -a\,y + b\,y^\alpha + c$ *in* (t_0, t_1);

(ii) $\displaystyle\int_{t_0}^{t_1} y(\tau)\,d\tau < \frac{\delta^2}{4c}$, $y(t_0) < \frac{\delta}{\sqrt{2}}$, *with* $\delta \in (0, (a/b)^{\frac{1}{\alpha-1}})$.

Then, necessarily

$$y(t) < \delta, \quad \text{for all } t \in [t_0, t_1).$$

Moreover, if $t_1 = \infty$ we have also

$$\lim_{t\to\infty} y(t) = 0.$$

We can now show the announced regularity property of weak solutions.

Proposition 3.3.6. *Let* s $\equiv (v, \omega, \gamma)$ *be a weak solution to (3.1), with $\mathscr{C}$ of class C^2, and initial data of finite energy, in the sense of* Proposition 3.3.3. *Then, there exists $t_0 = t_0(\mathsf{s}) > 0$ such that, setting $I_{t_0,T} = (t_0, t_0 + T)$,*

$$v \in C^0(\overline{I_{t_0,T}}; W_0^{1,2}(\mathscr{C})) \cap L^\infty(t_0, \infty; W_0^{1,2}(\mathscr{C})) \cap L^2(I_{t_0,T}; W^{2,2}(\mathscr{C})),$$

$$v_t \in L^2(I_{t_0,T}; H(\mathscr{C})), \quad \omega \in W^{1,\infty}(I_{t_0,T}), \quad \gamma \in W^{2,\infty}(I_{t_0,T}; \mathsf{S}^1),$$
(3.12)

for all $T > 0$. Moreover, there is $p \in L^2(I_{t_0,T}; W^{1,2}(\mathscr{C}))$, all $T > 0$, such that (v, p, ω, γ) satisfies $(3.1)_{1,2}$ a.e. in $\mathscr{C} \times (t_0, \infty)$. Finally,

$$\lim_{t\to\infty} \|v(t)\|_{1,2} = 0.$$
(3.13)

Proof. To begin with, we observe that, in view of the property (a) of weak solutions, for any given $\varepsilon, \eta > 0$ there exists $t_0 = t_0(\varepsilon, \eta, \mathsf{s}) > 0$ such that

$$\|\nabla v(t_0)\|_2 < \varepsilon, \quad \int_{t_0}^\infty \|\nabla v(\tau)\|_2^2 \, d\tau < \eta.$$
(3.14)

Actually, since $v \in L^2(0, \infty; W_0^{1,2}(\mathscr{C}))$, we can find an increasing, unbounded sequence $\{t_k\} \subset S$ such that, for any $\varepsilon > 0$, there is $\overline{k} \in \mathbb{N}$ satisfying

$$\|\nabla v(t_k)\|_2 < \varepsilon, \quad \text{for all } k \geq \overline{k}.$$

Moreover, by the same property, we infer that for any $\eta > 0$ there is $\overline{t} > 0$ such that

$$\int_{\overline{t}}^\infty \|\nabla v(\tau)\|_2^2 \, d\tau < \eta.$$

Thus, we get (3.14) by choosing $t_0 = t_{k^*}$ where $k^* \geq \overline{k}$, $t_{k^*} \geq \overline{t}$. We next construct a strong solution $\widetilde{\mathsf{s}} \equiv (\widetilde{v}, \widetilde{\omega}, \widetilde{\gamma})$ in the interval $[t_0, T^*)$, for *some* $T^* > t_0$. The existence of $\widetilde{\mathsf{s}}$ is accomplished by using, again, Galerkin method in conjunction with suitable "energy" estimates. Here, we shall limit ourselves to derive the latter, referring the reader to Chapter 4 in Mazzone (2012) for further technical details. To this aim, we dot-multiply both sides of $(3.1)_1$ by v_t and integrate by parts over $\mathscr{C}$. Using $(3.1)_2$ and $(3.1)_3$, we deduce that

$$\frac{\mu}{2}\frac{d}{dt}\|\nabla v\|_2^2 + \rho\|v_t\|_2^2 - C\dot{a}^2 = -\rho(v\cdot\nabla v, v_t) - 2\omega(e_3\times v, v_t). \quad (3.15)$$

Also, by dot-multiplying $(3.1)_1$ by $P\Delta v$, with P orthogonal projection of $L^2(\mathscr{C})$ onto $H(\mathscr{C})$ it follows that

$$\mu\|P\Delta v\|_2^2 = \rho\Big(\big[v_t + (\dot{a} + \beta^2\gamma_2)\,e_3\times x + v\cdot\nabla v + 2\omega\,e_3\times v\big], P\Delta v\Big). \quad (3.16)$$

By a straightforward use of the Schwarz inequality, from (3.16) we infer that

$$\|P\Delta v\|_2^2 \le c_1\left(\|v_t\|_2^2 + \|v\cdot\nabla v\|_2^2 + |\omega|^2\|v\|_2^2 + |\gamma_2|^2\right), \quad (3.17)$$

where, here and in the following, $c_i > 0$, $i\in\mathbb{N}$, denote constants depending at most on the physical parameters of the problem and the data, s_0, of the weak solution s at $t = 0$. Now, we notice that, since $|\gamma(t)| = 1$ for all $t \ge t_0$, from the strong energy inequality in (c) and (3.6)–(3.9) we can deduce that

$$\|v(t)\|_2 + |\omega(t)| \le c_2, \quad \text{all } t \ge t_0. \quad (3.18)$$

As a consequence, by using (3.18) on the right-hand side of (3.15) and (3.17) along with (3.9) and Cauchy-Schwarz inequality, we show that

$$\frac{d}{dt}\|\nabla v\|_2^2 + c_3\|v_t\|_2^2 \le c_4(\|v\cdot\nabla v\|_2^2 + 1),$$
$$\|P\Delta v\|_2^2 \le c_5(\|v_t\|_2^2 + \|v\cdot\nabla v\|_2^2 + 1). \quad (3.19)$$

Multiplying both sides of $(3.19)_2$ by $c_3/(2c_5)$ and summing side-by-side the resulting inequality and $(3.19)_1$, we get

$$\frac{d}{dt}\|\nabla v\|_2^2 + c_6\|v_t\|_2^2 + c_7\|P\Delta v\|_2^2 \le c_8\left(\|v\cdot\nabla v\|_2^2 + 1\right). \quad (3.20)$$

Employing the fact that $\mathscr{C}$ is of class C^2, we have (e.g. Theorem IV.6.1 in Galdi, 2011)

$$\|v\|_{2,2}^2 \le c_9\|P\Delta v\|_2^2, \quad (3.21)$$

whereas, by classical embedding theorems and Cauchy–Schwarz inequality, we show for arbitrary $\lambda > 0$

$$\|v\cdot\nabla v\|_2^2 \le \|v\|_\infty^2\|\nabla v\|_2^2 \le c_{10}\|\nabla v\|_2^3\|v\|_{2,2} \le c_{11}\|\nabla v\|_2^6 + \lambda\|v\|_{2,2}^2. \quad (3.22)$$

where $c_{11} \to \infty$ as $\lambda \to 0$. With the choice $\lambda = c_7/(2c_8c_9)$, from (3.20)–(3.22) we thus conclude

$$\frac{d}{dt}\|\nabla \boldsymbol{v}\|_2^2 + c_6\|\boldsymbol{v}_t\|_2^2 + c_{12}\|\boldsymbol{v}\|_{2,2}^2 \leq c_{13}\left(\|\nabla \boldsymbol{v}\|_2^6 + 1\right). \qquad (3.23)$$

From (3.23) we derive, in particular,

$$\frac{dz}{dt} \leq c_{14}\, z^3, \qquad (3.24)$$

with $z := \|\nabla \boldsymbol{v}\|_2^2 + 1$. By integrating this differential inequality and using (3.23), we can guarantee that there exist some interval $[t_0, t_0 + T^*)$ with $T^* \geq c_{15}/\|\nabla \boldsymbol{v}(t_0)\|_2^4$, and continuous functions G and G_1 defined on $[t_0, t_0 + T^*)$ such that

$$\|\boldsymbol{v}(t)\|_{1,2} \leq G(t), \quad \int_{t_0}^t \left(\|\boldsymbol{v}_\tau(\tau)\|^2 + \|\boldsymbol{v}(\tau)\|_{2,2}^2\right) d\tau \leq G_1(t). \qquad (3.25)$$

We may then combine the latter with the classical Galerkin method, and interpolation inequalities (Lions & Magenes, 1971, Chapter 1, Theorem 3.1), to show the existence of a solution $\widetilde{\mathsf{s}} \equiv (\widetilde{\boldsymbol{v}}, \widetilde{\omega}, \widetilde{\boldsymbol{\gamma}})$ corresponding to the initial data $(\boldsymbol{v}(t_0), \omega(t_0), \boldsymbol{\gamma}(t_0))$, where, we recall, $(\boldsymbol{v}, \omega, \boldsymbol{\gamma}) \equiv \mathsf{s}$ is the given weak solution, and such that, setting $I_{t_0,\tau} := (t_0, t_0 + \tau)$,

$$\widetilde{\boldsymbol{v}} \in C^0(\overline{I_{t_0,\tau}}; W_0^{1,2}(\mathscr{C})) \cap L^\infty(I_{t_0,\tau}; W_0^{1,2}(\mathscr{C})) \cap L^2(I_{t_0,\tau}; W^{2,2}(\mathscr{C})),$$

$$\widetilde{\boldsymbol{v}}_t \in L^2(I_{t_0,\tau}; H(\mathscr{C})), \quad \widetilde{\omega} \in W^{1,\infty}(I_{t_0,\tau}), \quad \widetilde{\boldsymbol{\gamma}} \in W^{2,\infty}(I_{t_0,\tau}; \mathbb{S}^1),$$

$$\text{for all } \tau \in (0, T^*), \qquad (3.26)$$

with t_0 satisfying (3.14) for appropriate ε and η to be chosen shortly. Since, as immediately checked by the Sobolev embedding theorem, $\widetilde{\boldsymbol{v}}$ satisfies (3.11) for suitable $r > 3$, by Proposition 3.3.4 it follows that $\widetilde{\mathsf{s}} = \mathsf{s}$ on the interval $[t_0, t_0 + T^*)$. Next, by proceeding exactly as in the proof of Proposition 2 in Galdi et al. (2014), one shows that if $T^* < \infty$ then necessarily

$$\lim_{t \to T^{*-}} \|\nabla \boldsymbol{v}(t)\|_2 = \infty. \qquad (3.27)$$

However, this condition cannot hold. Actually, from (3.23) we see that $\|\nabla \boldsymbol{v}\|_2^2$ satisfies assumption (i) of Lemma 3.3.5 with $a = c_{12}$ and $b = c = c_{13}$. Moreover, we choose ε and η in (3.14) in a such a way that also assumption (ii) is satisfied. As a consequence, Lemma 3.3.5 implies that (3.27) cannot occur. So, $\|\nabla \boldsymbol{v}(t)\|_2 \in L^\infty(t_0, \infty)$ and $(\boldsymbol{v}, \omega, \boldsymbol{\gamma})$ satisfies

(3.12) for all $T > 0$. Furthermore, again by Lemma 3.3.5 we also infer the validity of (3.13), which, along with the classical Poincaré inequality:

$$\|w\|_2 \leq \kappa \|\nabla w\|_2, \quad w \in W_0^{1,2}(\mathscr{C}), \tag{3.28}$$

concludes the proof of the proposition. $\square$

From Proposition 3.3.4 and Proposition 3.3.6 we at once deduce the following important consequence.

Corollary 3.3.7. *Let* $\mathsf{s} \equiv (\boldsymbol{v}, \omega, \boldsymbol{\gamma})$ *be a weak solution to* (3.1) *with* $\mathscr{C}$ *of class* C^2. *Then there exists* $t_0 > 0$ *such that for all* $t \geq t_0$ *the following properties hold:*

  (i) s *is unique in the class of weak solutions;*
  (ii) s *depends continuously upon the data at any* $\tau \geq t_0$, *in the class of weak solutions in the sense specified by Proposition 3.3.4.*

3.4 On the asymptotic behaviour of weak solutions

In the previous section we have established, among other things, that in any motion described by a weak solution emerging from initial data of finite kinetic energy, the velocity of the liquid must eventually tend to zero for large times in appropriate norm (see (3.13)). In the present one, we shall investigate the asymptotic behaviour of the angular velocity, ω, and the orientation, $\boldsymbol{\gamma}$, of the rigid body, thus providing a complete description of the asymptotic behaviour of the *coupled* system liquid-body. To this end, we borrow some tools from the classical theory of dynamical systems that, however, must be appropriately adapted to the present situation.

Let $\mathsf{s} \equiv (\boldsymbol{v}, \omega, \boldsymbol{\gamma})$ be a given weak solution to (3.1). We define the Ω-limit set of s:

$$\Omega(\mathsf{s}) = \Big\{ (\boldsymbol{u}, r, \boldsymbol{q}) \in H(\mathscr{C}) \times \mathbb{R} \times \mathsf{S}^1 : \text{there are } t_k \geq 0 \text{ with } t_k \to \infty \text{ s.t.}$$
$$\lim_{k\to\infty} \|\boldsymbol{v}(t_k) - \boldsymbol{u}\|_2 = \lim_{k\to\infty} |\omega(t_k) - r| = \lim_{k\to\infty} |\boldsymbol{\gamma}(t_k) - \boldsymbol{q}| = 0 \Big\}. \tag{3.29}$$

Since we do not know whether weak solutions are unique, $\Omega(\mathsf{s})$ may not depend just on the initial data of s, but, rather on s itself.

Our objective is to characterize the structure of $\Omega(\mathsf{s})$. To this end, set $\mathscr{H} := H(\mathscr{C}) \times \mathbb{R} \times \mathsf{S}^1$, endowed with its natural topology. For $t \geq 0$, we shall denote by $w(t; z)$ a weak solution to (3.1) corresponding to the initial data $z \in \mathscr{H}$, in the sense of Proposition 3.3.3.

Definition 3.4.1. $\Omega(s)$ is positively invariant if

$$y \in \Omega(s) \implies w(t; y) \in \Omega(s), \quad \text{all } t \geq 0,$$

for all weak solutions $w(t; y)$.

The next result ensures the invariance property of $\Omega(s)$, provided that $s(t; s_0)$ is "asymptotically regular". Even though the proof is quite standard, we wish to include it nevertheless, for the reader's sake.

Proposition 3.4.2. *Let $s(t; s_0)$ be a weak solution to (3.1). Suppose there is $t_0 > 0$ such that the following properties hold.*

(i) Asymptotic Uniqueness:

$$s(t + \tau; s_0) = s(t; s(\tau; s_0)), \quad \text{all } \tau \geq t_0, \text{ and } t \geq 0;$$

(ii) Asymptotic Continuous Data Dependence:

$$\{t_k\} \subset [t_0, \infty) \text{ with } s(t_k; s_0) \to y \text{ in } \mathscr{H}$$
$$\implies s(t; s(t_k; s_0)) \to w(t; y) \text{ in } \mathscr{H}, \text{ all } t \geq 0.$$

Then $\Omega(s)$ is positively invariant.

Proof. Let $y \in \Omega(s)$ and let $w(t; y)$ be a corresponding weak solution. We have to show that for each $t \geq 0$ there is $\{\tau_n\} \subset \mathbb{R}_+$ unbounded and such that

$$s(\tau_n; s_0) \to w(t; y) \text{ in } \mathscr{H}. \qquad (3.30)$$

We observe that, by definition,

$$s(t_n; s_0) \to y \text{ in } \mathscr{H}, \qquad (3.31)$$

for some unbounded sequence $\{t_n\} \subset \mathbb{R}_+$. Now, let $\bar{n}$ be such that $t_n \geq t_0$, for all $n \geq \bar{n}$ and set $\tau_n := t + t_n$, for all $n \geq \bar{n}$, and $t \geq 0$. By (i) we thus have

$$s(\tau_n; s_0) := s(t_n + t; s_0) = s(t; s(t_n; s_0)), \qquad (3.32)$$

whereas, by (ii) and (3.31) we also have

$$s(t; s(t_n; s_0)) \to w(t; y) \text{ in } \mathscr{H}.$$

Consequently, (3.30) follows from the latter and (3.32).   □

We are now in a position to give the following characterization of the Ω-limit set of any weak solution to (3.1)

Proposition 3.4.3. *Let* $\mathsf{s} \equiv (\boldsymbol{v}, \omega, \boldsymbol{\gamma})$ *be a weak solution to* (3.1)*, with* $\mathscr{C}$ *of class* C^2*, and initial data of finite energy in the sense of Proposition 3.3.3. Then, the corresponding* Ω-*limit set defined in* (3.29) *admits the following characterization: either*

$$\Omega(\mathsf{s}) = \big\{(\mathbf{0}, 0, \boldsymbol{e}_1)\big\},$$

or

$$\Omega(\mathsf{s}) = \big\{(\mathbf{0}, 0, -\boldsymbol{e}_1)\big\}.$$

Proof. In view of Corollary 3.3.7 and Proposition 3.4.2 we deduce that $\Omega(\mathsf{s})$ is invariant in the class of weak solutions. Therefore, from (3.13) and (3.1) we infer that the dynamics on $\Omega(\mathsf{s})$ is governed by the following set of equations

$$\boldsymbol{v} \equiv \mathbf{0}, \quad \beta^2 \gamma_2 \, \boldsymbol{e}_3 \times \boldsymbol{x} = -\nabla p, \quad \dot{\omega} = \beta^2 \gamma_2, \quad \dot{\boldsymbol{\gamma}} + \omega \, \boldsymbol{e}_3 \times \boldsymbol{\gamma} = \mathbf{0}. \quad (3.33)$$

Operating with curl on both sides of the second equation in (3.33) we find $\gamma_2 \equiv 0$, so that the third equation furnishes $\omega \equiv \overline{\omega}$, for some $\overline{\omega} \in \mathbb{R}$, which once replaced into the last equation in (3.33) shows $\overline{\omega} \gamma_1 = 0$. Taking into account that $|\boldsymbol{\gamma}| \equiv 1$, from all the above we then conclude $\overline{\omega} = 0$ and $\gamma_1 = \pm 1$. Consequently,

$$\Omega(\mathsf{s}) \subset \big\{(\mathbf{0}, 0, \boldsymbol{e}_1)\big\} \cup \big\{(\mathbf{0}, 0, -\boldsymbol{e}_1)\big\}.$$

However, the weak solution s must have, in particular, $\boldsymbol{\gamma} \in C^0([0, \infty); \mathsf{S}^1)$ (see (b) in Definition 3.3.2). This implies that $\Omega(\mathsf{s})$ has to be connected, and the proof of the proposition is therefore completed. $\qquad\square$

We are now ready to give a complete description of the asymptotic behaviour of weak solutions to (3.1).

Theorem 3.4.4. *Let* $(\boldsymbol{v}, \omega, \boldsymbol{\gamma})$ *be a weak solution to* (3.1) *with* $\mathscr{C}$ *of class* C^2*, and initial data of finite energy in the sense of Proposition 3.3.3. Then,*

$$\lim_{t \to \infty} \big(\|\boldsymbol{v}(t)\|_{2,2} + \|\boldsymbol{v}_t(t)\|_2 \big) = 0, \quad (3.34)$$

so that, in particular,

$$\lim_{t \to \infty} \big(\max_{\boldsymbol{x} \in \mathscr{C}} |\boldsymbol{v}(\boldsymbol{x}, t)| \big) = 0. \quad (3.35)$$

Moreover,

$$\lim_{t\to\infty} |\omega(t)| = 0, \quad \lim_{t\to\infty} |\gamma(t) - \alpha\,e_1| = 0, \qquad (3.36)$$

where $\alpha = 1$ or $\alpha = -1$.

Proof. We commence by observing that, as a result of the classical embedding inequality

$$\max_{x\in\mathscr{C}} |w(x)| \le C_{17}\|w\|_{2,2}, \quad w \in W^{2,2}(\mathscr{C}),$$

property (3.35) follows from (3.34). Thus, in view of this and Proposition 3.4.2, to prove the theorem completely we only have to prove the validity of (3.34). This can be achieved by the following procedure. By virtue of Proposition 3.3.6, our weak solution must have $v(t) \in W^{2,2}(\mathscr{C})$, for a.e. $t \in [t_0, \infty)$. For simplicity, and without loss of generality, we assume that

$$v(t_0) \in W^{2,2}(\mathscr{C}). \qquad (3.37)$$

Next, we formally take the time derivative of both sides of $(3.1)_1$, dot-multiply both sides of the resulting equation by v_t and integrate by parts over $\mathscr{C}$. By taking into account $(3.1)_{2,3,5}$ and (3.2), we easily show that $(\nu := \mu/\rho)$

$$\frac{1}{2}\frac{d}{dt}E_1 = -C\,\beta^2\dot{a}\,\omega\,\gamma_1 - 2\dot{\omega}\,(e_3\times v, v_t) - (v_t\cdot\nabla v, v_t) - \nu\|\nabla v_t\|_2^2, \quad (3.38)$$

where

$$E_1 := \|v_t\|_2^2 - \frac{C}{\rho}\dot{a}^2.$$

Notice that, by (3.9),

$$c_0\,\|v_t\|_2^2 \le E_1 \le \|v_t\|_2^2. \qquad (3.39)$$

Using the Cauchy–Schwarz inequality and the Poincaré inequality (3.28), and recalling that $|\gamma| = 1$, we deduce that

$$\frac{d}{dt}E_1 + C_1\|\nabla v_t\|_2^2 \le C_2\left(-(v_t\cdot\nabla v, v_t) + \omega^2 + \|v\|_2^2 + \|v\|_2\,\|v_t\|_2^2\right). \qquad (3.40)$$

Here and in the rest of the proof, C_i, $i \in \mathbb{N}$, denotes a positive constant depending, at most, on the physical parameters of the problem and the data of the given weak solution at $t = 0$. Applying Hölder's inequality,

Pendulum with a cavity entirely filled with a viscous liquid 51

the Sobolev embedding theorem, and Young's inequality, in that order, we infer that

$$|(\boldsymbol{v}_t \cdot \nabla \boldsymbol{v}, \boldsymbol{v}_t)| \leq \|\boldsymbol{v}_t\|_4^2 \|\nabla \boldsymbol{v}\|_2 \leq C_3 \|\nabla \boldsymbol{v}_t\|_2^{\frac{3}{2}} \|\boldsymbol{v}_t\|_2^{\frac{1}{2}} \|\nabla \boldsymbol{v}\|_2$$
$$\leq \tfrac{1}{2} C_1 \|\nabla \boldsymbol{v}_t\|_2^2 + C_4 \left(\|\boldsymbol{v}_t\|_2^6 + \|\nabla \boldsymbol{v}\|_2^6 \right). \tag{3.41}$$

Taking into account (3.18) we may combine the latter with (3.40), (3.41) to obtain

$$\frac{\mathrm{d}}{\mathrm{d}t} E_1 + \tfrac{1}{2} C_1 \|\nabla \boldsymbol{v}_t\|_2^2 \leq C_5 + C_6 \left(\|\boldsymbol{v}_t\|_2^6 + \|\nabla \boldsymbol{v}\|_2^6 \right). \tag{3.42}$$

From (3.42), (3.39), and (3.23) to show, in particular, the validity of (3.24) with $z := \|\nabla \boldsymbol{v}\|_2^2 + E_1 + 1$. Integrating the differential inequality thus obtained and using again (3.23), (3.39), and (3.42), we prove, in addition to the bounds (3.25), that

$$\|\boldsymbol{v}_t(t)\|_2 \leq G_2(t), \quad \int_{t_0}^t \|\nabla \boldsymbol{v}_\tau(\tau)\|^2 \, \mathrm{d}\tau \leq G_3(t),$$

with G_i, $i = 2, 3$, continuous functions in the interval $[t_0, t_0 + T^*)$, where $T^* \geq C_7/(\|\nabla \boldsymbol{v}(t_0)\|_2^4 + \|\boldsymbol{v}_t(t_0)\|_2^4 + 1)$, and $C_7 > 0$ independent of t_0. We now go back to (3.1)$_1$, dot-multiply both sides by $\boldsymbol{v}_t$ and integrate over $\mathscr{C}$. We obtain

$$\rho \|\boldsymbol{v}_t\|_2^2 - C \dot{a}^2 = \frac{C \beta^2}{\rho} \gamma_2 \dot{a} - \rho (\boldsymbol{v} \cdot \nabla \boldsymbol{v}, \boldsymbol{v}_t) - 2\omega(e_3 \times \boldsymbol{v}, \boldsymbol{v}_t) + \mu(\Delta \boldsymbol{v}, \boldsymbol{v}_t).$$

Now using the Schwarz inequality, (3.39), (3.18), and (3.22) we can show (formally)

$$\|\boldsymbol{v}_t(t_0)\|_2 \leq C_8(\|\boldsymbol{v}(t_0)\|_{2,2}^3 + \|\boldsymbol{v}(t_0)\|_{2,2} + 1), \tag{3.43}$$

which implies, on the one hand, by (3.37) that $\|\boldsymbol{v}_t(t_0)\|_2$ is well-defined, and, on the other hand, that $T^* \geq C_9/[D(\|\boldsymbol{v}(t_0)\|_{2,2}) + 1]$ where $D = D(\sigma)$ is a polynomial satisfying $D(0) = 0$. Collecting all the above information, we may thus employ the standard Galerkin method and show the existence of a solution $(\tilde{\boldsymbol{v}}, \tilde{\omega}, \tilde{\gamma})$ with data $(\boldsymbol{v}(t_0), \omega(t_0), \gamma(t_0))$ that, besides (3.26), satisfies also

$$\tilde{\boldsymbol{v}}_t \in L^\infty(t_0, t_0 + \tau; H(\mathscr{C})) \cap L^2(t_0, t_0 + \tau; W_0^{1,2}(\mathscr{C})), \quad \text{all } \tau \in (0, T^*);$$

see, for example, Chapter 4 in Mazzone (2012) or Chapter 6.3 in Ladyzhenskaya (1969) for technical details. However, by the uniqueness property of Proposition 3.3.4, this solution must coincide with the given weak solution on $[t_0, t_0 + T^*)$. We can then show that in fact $T^* = \infty$.

Actually, if $T^* < \infty$, it easily follows that necessarily $\|v(t)\|_{2,2}$ must become unbounded in a left-neighbourhood of $t_0 + T^*$. Let us show that such a situation cannot occur. To this end, we begin to observe that, by what just shown, the given weak solution satisfies (3.40) in $(t_0, t_0 + T^*)$. Now, by Hölder's inequality, the Sobolev embedding theorem, and Young's inequality, in the order, we obtain

$$|(v_t \cdot \nabla \quad v, v_t)| = |(v_t \cdot \nabla v_t, v)| \leq \|v_t\|_4 \|v\|_4 \|\nabla v_t\|_2$$

$$\leq C_{10} \|\nabla v_t\|_2^{\frac{7}{4}} \|v_t\|_2^{\frac{1}{4}} \|\nabla v\|_2 \leq \tfrac{1}{2} C_1 \|\nabla v_t\|_2^2 + C_{11} \|v_t\|_2^2 \|\nabla v\|_2^8.$$

Using this inequality in (3.40) we deduce that

$$\frac{d}{dt} E_1 + \tfrac{1}{2} C_1 \|\nabla v_t\|_2^2 \leq C_{12} \left[\omega^2 + \|v\|_2^2 + (\|v\|_2 + \|\nabla v\|_2^8) \|v_t\|_2^2\right], \quad (3.44)$$

However, v and ω must satisfy (3.12), so that, integrating both sides of (3.44) over $(t_0, t_0 + T^*)$ and using (3.37), (3.39), and (3.43) we infer that

$$v_t \in L^\infty(t_0, t_0 + T^*; H(\mathscr{C})). \quad (3.45)$$

Furthermore, as a consequence of (3.17), (3.21), and (3.22) with $\lambda = 1/(2c_1 c_9)$, we have

$$\|v(t)\|_{2,2} \leq C_{13} (\|\nabla v(t)\|_2^3 + \|v_t(t)\|_2 + |\omega(t)| \|v(t)\|_2 + |\gamma_2(t)|),$$

$$\text{for a.e. } t \in [t_0, t_0 + T^*). \quad (3.46)$$

Therefore, this inequality along with (3.45), (3.12) and the fact that $|\gamma| = 1$ allows us to conclude $v \in L^\infty(t_0, t_0 + T^*; W^{2,2}(\mathscr{C}))$, which, in turn, implies $T^* = \infty$. We now go back to (3.44) –valid for all $t \in (t_0, \infty)$– and use (3.28) to show that

$$\frac{d}{dt} E_1 + C_{14} \left[1 - (\|v\|_2 + \|\nabla v\|_2^8)\right] \|v_t\|_2^2 \leq C_{12} \left(\omega^2 + \|v\|_2^2\right). \quad (3.47)$$

By (3.13) we may find $t_1 \geq t_0$ such that

$$C_{14} (1 - \|v\|_2 - \|\nabla v\|_2^8) \geq C_{15}, \quad \text{for all } t \geq t_1,$$

which, once replaced into (3.47), with the help of (3.39) yields

$$\frac{d}{dt} E_1 + C_{15} E_1 \leq F(t), \quad (3.48)$$

where $F(t) := C_{12} \left(\omega^2(t) + \|v(t)\|_2^2\right)$. Notice that, by Proposition 3.3.6 and Proposition 3.4.3

$$\lim_{t \to \infty} F(t) = 0. \quad (3.49)$$

By using a Gronwall-like lemma in (3.48), again thanks to (3.39) it follows that[1] for all $t \geq 2t_1$

$$\|\boldsymbol{v}_t(t)\|_2^2 \leq C_{16}\left(\|\boldsymbol{v}_t(t/2)\|_2^2 \, \mathrm{e}^{-C_{15}t/2} + \int_{t/2}^t \mathrm{e}^{-C_{15}(t-s)} F(s)\,\mathrm{d}s\right).$$

Using (3.45) (valid with $T^* = \infty$) and (3.49) we then show that for any $\varepsilon > 0$ there is $\bar{t} > 0$ such that

$$\|\boldsymbol{v}_t(t)\|_2^2 \leq C_{17}\,\mathrm{e}^{-C_{15}t/2} + \varepsilon\,\frac{C_{16}}{C_{15}}, \quad \text{for all } t \geq \bar{t},$$

namely,

$$\lim_{t\to\infty} \|\boldsymbol{v}_t(t)\|_2 = 0. \tag{3.50}$$

As a result, (3.34) follows from (3.13), (3.46) (valid with $T^* = \infty$), (3.36), and (3.50). This concludes the proof of the theorem. $\qquad\square$

Remark 3.4.5. From Theorem 3.4.4 and $(3.1)_{4,5}$ it also follows that

$$\lim_{t\to\infty} |\dot{\omega}(t)| = \lim_{t\to\infty} |\dot{\boldsymbol{\gamma}}(t)| = 0.$$

3.5 Attainability and stability of the equilibrium configurations

The results proved in Theorem 3.4.4 imply that the coupled solid-liquid system $\mathscr{S}$ will eventually reach an equilibrium configuration where the liquid is at rest, and the centre of mass G of $\mathscr{S}$ is on the vertical axis passing through the point of suspension O. However, the theorem does not specify whether G lies above O (i.e. $\boldsymbol{\gamma} = -\boldsymbol{e}_1$), or below O (i.e. $\boldsymbol{\gamma} = \boldsymbol{e}_1$). The objective of this section is to show that, under suitable conditions on the initial data, $\mathscr{S}$ will reach the equilibrium configuration where G is in its lowest position (i.e. $\boldsymbol{\gamma} = \boldsymbol{e}_1$). It is worth observing that if $\mathscr{S}$ is initially released from rest then the conditions of the theorem are certainly satisfied.

More specifically, we have the following.

Theorem 3.5.1. *Let $\mathscr{C}$ be of class C^2, and let $(\boldsymbol{v}_0, \omega_0, \boldsymbol{\gamma}_0) \in H(\mathscr{C}) \times \mathbb{R} \times \mathsf{S}^1$ be given with*

$$\rho\,\|\boldsymbol{v}_0\|_2^2 + C\,(\omega_0 - a(0))^2 < 2\,C\,\beta^2\,(1 + \gamma_{1,0}). \tag{3.51}$$

---

[1]Observe that by (3.38) and the property just shown, it follows that the function $t \mapsto \|\boldsymbol{v}_t(t)\|_2$ is absolutely continuous for all "large" t.

Then all weak solutions corresponding to initial data $(\boldsymbol{v}_0, \omega_0, \boldsymbol{\gamma}_0)$ tend to the equilibrium configuration $(\boldsymbol{v} \equiv \boldsymbol{0}, \omega \equiv 0, \boldsymbol{\gamma} \equiv \boldsymbol{e}_1)$, namely, the one where the centre of mass lies in its lowest position.

Proof. Suppose, by contradiction, that the final equilibrium position is, instead, $(\boldsymbol{v} \equiv \boldsymbol{0}, \omega \equiv 0, \boldsymbol{\gamma} \equiv -\boldsymbol{e}_1)$. Then, passing to the limit $t \to \infty$ in the energy inequality (3.10) with $s = 0$, and taking into account Theorem 3.4.4, we find, in particular,

$$C\beta^2 + \mu \int_0^\infty \|\nabla \boldsymbol{v}(t)\|_2^2 dt \leq \tfrac{1}{2}\left(\rho\|\boldsymbol{v}_0\|_2^2 + C(\omega_0 - a(0))^2 - 2C\beta^2\gamma_{1,0}\right),$$

which cannot be true whenever the initial data satisfy (3.51). $\square$

With the help of the previous result, we can prove the following theorem.

Theorem 3.5.2. *Suppose $\mathscr{C}$ of class C^2. Then the equilibrium configuration $\mathcal{C}_1 := (\boldsymbol{v} \equiv \boldsymbol{0}, \omega \equiv 0, \boldsymbol{\gamma} \equiv -\boldsymbol{e}_1)$, namely, the one where the centre of mass lies in its highest position, is unstable in the sense of Lyapunov in the class of weak solutions, whereas the configuration $\mathcal{C}_2 := (\boldsymbol{v} \equiv \boldsymbol{0}, \omega \equiv 0, \boldsymbol{\gamma} \equiv \boldsymbol{e}_1)$, where the centre of mass lies in its lowest position is stable.*

Proof. Consider a weak solution corresponding to the initial data $\boldsymbol{v}(0) = \boldsymbol{0}$, $\omega(0) = 0$ and $\boldsymbol{\gamma}(0) = -\cos\delta\, \boldsymbol{e}_1 + \sin\delta\, \boldsymbol{e}_2$, $\delta \neq 0$. Since these data satisfy (3.51), any corresponding weak solution will tend to the equilibrium $(\boldsymbol{v} \equiv \boldsymbol{0}, \omega \equiv 0, \boldsymbol{\gamma} \equiv \boldsymbol{e}_1)$, no matter how close δ to zero, namely, no matter how close the initial conditions to the configuration $\mathcal{C}_1$. This shows the claimed instability property. Next, let $(\boldsymbol{v}, \omega, \boldsymbol{\gamma}')$ denote a perturbation to the configuration $\mathcal{C}_2$ in the class of weak solutions. This means that $(\boldsymbol{v}, \omega, \boldsymbol{e}_1 + \boldsymbol{\gamma}')$ is a weak solution to (3.1) corresponding to initial data, say, $(\boldsymbol{v}_0, \omega_0, \boldsymbol{e}_1 + \boldsymbol{\gamma}_0')$. From the strong energy inequality (3.10) and (3.8) we at once deduce that, for all $t \geq 0$,

$$c_0\left[\rho\|\boldsymbol{v}(t)\|_2^2 + C(\omega(t) - a(t))^2\right] - 2C\beta^2\gamma_1'(t)$$
$$\leq \left[\rho\|\boldsymbol{v}_0\|_2^2 + C(\omega_0 - a(0))^2\right] - 2C\beta^2\gamma_{10}'. \tag{3.52}$$

Moreover, from the condition $|\boldsymbol{e}_1 + \boldsymbol{\gamma}'(t)|^2 = 1$, all $t \geq 0$, we find

$$-2\gamma_1'(t) = (\gamma_1'(t))^2 + (\gamma_2'(t))^2, \quad \text{all } t \geq 0. \tag{3.53}$$

From (3.52) and (3.53), and recalling (3.2), we immediately deduce that

C_2 is stable in the sense of Lyapunov, namely, for any given $\varepsilon > 0$ there is $\delta(\varepsilon) > 0$ such that

$$\|v_0\|_2 + |\omega_0| + |\gamma_0'| < \delta \implies \|v(t)\|_2 + |\omega(t)| + |\gamma'(t)| < \varepsilon, \quad \text{for all } t > 0.$$

The proof of the theorem is completed. □

3.6 Conclusions and final remarks

We have shown that a physical pendulum containing an interior cavity entirely filled with a viscous (Navier–Stokes) liquid must eventually go to an equilibrium state where the liquid is at rest and the centre of mass of the system occupies its highest (configuration C_1) or lowest (C_2) position. Moreover, we have proved that the former is unstable, while the latter is stable, and also attainable provided the initial data satisfy (3.51).

The following two interesting questions are, however, left open.

In the first place, we do not know the rate at which the equilibrium configuration C_2 will be reached, at least for sufficiently large times. In fact, in analogy with similar problems of rigid bodies with a liquid-filled cavity, it is expected that the motion would be "chaotic" for some interval of time, but then, once the velocity of the liquid becomes "sufficiently small" (the latter, all other parameters kept fixed, depending on the magnitude of the viscosity), it is conjectured that the system should go to the equilibrium configuration at a very fast rate, possibly, even exponentially.

The second open question regards whether condition (3.51) on the initial data is indeed necessary for the proof of attainability of the equilibrium configuration C_2. Actually, given the instability property of C_1, we conjecture that C_2 should be reached from "almost all" initial data (of finite energy).

Acknowledgements

Work partially supported by NSF DMS Grant-1311983.

References

Alfriend, K.T. & Spencer, T.M. (1983) Comparison of filled and partly filled nutation dampers. *J. Astronaut. Sci.* **31**, no.2.

Bhuta, P.G. & Koval, L.R. (1966) A viscous ring damper for a freely precessing satellite. *Intern. J. Mech. Sci.* **8**, 383–395.

Boyevkin, V.I., Gurevich, Y.G., & Pavlov, Y.I. (1976) *Orientation of artificial satellites in gravitational and magnetic fields.* Nauka, Moscow (in Russian).

Chernousko, F.L. (1972) Motion of a rigid body with cavities containing a viscous Fluid, 1968, Moscow; NASA Technical Translations, May 1972.

Galdi, G.P. (2011) *An introduction to the mathematical theory of the Navier–Stokes equations. Steady-state problems.* Springer Monographs in Mathematics, Springer, New York.

Galdi, G.P., Mazzone, G., & Zunino, P. (2013) Inertial motions of a rigid body with a cavity filled with a viscous liquid. *Comptes Rendus Mécanique* **341**, 760–765.

Galdi, G.P., Mazzone, G., & Zunino P. (preprint) Inertial motions of a rigid body with a cavity filled with a viscous liquid, arXiv:1405.6596.

Kopachevsky, N.D. & Krein, S.G. (2000) *Operator approach to linear problems of hydrodynamics, Vol. 2: Nonself-adjoint problems for viscous fluids,* Birkhäuser Verlag, Basel-Boston-Berlin.

Krasnoshchekov, P.S. (1963) On oscillations of a physical pendulum having cavities filled with a viscous liquid. *J. Appl. Math. Mech.* **27**, 289–302.

Ladyzhenskaya, O.A. (1969) *The mathematical theory of viscous incompressible flow.* Revised second English edition. Gordon and Breach Science Publishers, New York-London.

Leung, A.Y.T. & Kuang, J.L. (2007) Chaotic rotations of a liquid-filled solid. *J. Sound and Vibr.* **302**, 540–563.

Lions, J.L. & Magenes, E. (1971) *Non-homogeneous boundary value problems and applications.* Die Grundlehren der mathematischen Wissenschaften, **181**, Springer-Verlag Berlin/Heidelberg/New York.

Mazzone, G. (2012) *A mathematical analysis of the motion of a rigid body with a cavity containing a Newtonian fluid.* PhD Thesis, Department of Mathematics, Università del Salento.

Moiseyev, N.N. & Rumyantsev, V.V. (1968) *Dynamic stability of bodies containing fluid.* Springer-Verlag, New York.

Pivovarov, M.L. & Chernousko, F.L. (1990) Oscillations of a rigid body with a toroidal cavity filled with a viscous liquid. *Prikl. Mat. Mekh.* **54**, 164–168.

Pook, P.L. (2011) *Understanding pendulums. A brief introduction.* Springer Science & Business Media.

Sarychev, V.A. (1978) Problems of the orientation of artificial satellites. *Itogi Nauki i Tekhniki. Issledovanie kosmicheskogo prostranstva* **11**, Viniti, Moscow (in Russian).

Silvestre, A.L. & Takahashi, T. (2012) On the motion of a rigid body with a cavity filled with a viscous liquid. *Proc. Roy. Soc. Edinburgh Sect. A* **142**, 391–423.

Smirnova, E.P. (1975) The motion of a liquid of high viscosity in a rotating torus. *Prikl. Mat. Mekh.* **39** no. 1.

4

Modal dependency and nonlinear depletion in the three-dimensional Navier–Stokes equations

John D. Gibbon

Department of Mathematics,
Imperial College London,
London SW7 2AZ, UK.
j.d.gibbon@ic.ac.uk   www2.imperial.ac.uk/~jdg

Abstract

A summary is given of recent work on numerical and analytical studies in nonlinear depletion in the 3D Navier–Stokes equations by Donzis et al. (2013) and Gibbon et al. (2014). These results are specifically discussed in terms of modal dependency, where the high modes are controlled by the lowest modes. The modes (frequencies) used are L^{2m}-norms of vorticity.

4.1 Introduction

In fluid dynamics the idea that the low modes of a system might conceivably control the high modes has long been attractive, not least because it suggests a potential mechanism for the reduction in size and cost of computing high dimensional systems, in particular three dimensional turbulent flows. The idea has emerged in different guises over the last 40 years including the development of centre manifolds for ODEs (Guckenheimer & Holmes, 1997) and inertial manifolds for PDEs (Foias, Sell, & Temam, 1988, Titi, 1990, Foias & Titi, 1991, Robinson, 1996), together with the concept of determining modes and nodes (Foias & Temam, 1984). Success has been limited to the construction of an inertial manifold for the one-dimensional Kuramoto–Sivashinsky equation (Foias, Jolly, Kevrekidis, Sell, & Titi, 1988) and the achievement of an estimate for the number of determining modes and nodes for the 2D Navier–Stokes equations (Foias & Titi, 1991, Jones & Titi, 1993, Olson & Titi, 2003, 2008, Farhat, Jolly, & Titi, 2015). The 3D Navier–Stokes

Published in *Recent Progress in the Theory of the Euler and Navier-Stokes Equations*, edited by James C. Robinson, José L. Rodrigo, Witold Sadowski, & Alejandro Vidal-López. ©Cambridge University Press 2016.

equations, however, have remained stubbornly resistant, the main obstacle being the issue of existence and uniqueness of solutions (Constantin & Foias, 1988, Doering & Gibbon, 1995, Foias, Manley, Rosa, & Temam, 2001, Escauriaza, Seregin, & Šverák, 2003).

Another major feature of work on the 3D Navier–Stokes equations has been the difficulty in interpreting the results of large scale numerical simulations in the context of the highly limited analytical results that are available: for early numerical results see Orszag & Patterson (1972), Rogallo (1981), Kerr (1985), Eswaran & Pope (1988), Jimenez, Wray, Saffman, & Rogallo (1993), Moin & Mahesh (1998). However, now that higher resolution is available, it is perhaps time to renew the idea of interpreting some more recent simulations to see if they can inform the analysis by suggesting new ways of considering the Navier–Stokes regularity problem: for more recent numerical work see Kurien & Taylor (2005), Ishihara, Gotoh, & Kaneda (2009), Donzis, Yeung, & Sreenivasan (2008), Donzis & Yeung (2010), Yeung, Donzis, & Sreenivasan (2012), Kerr (2012, 2013) and other data-bases (Johns Hopkins, 2014).

In recent numerical experiments described in Donzis, Gibbon, Gupta, Kerr, Pandit, & Vincenzi (2013), a set of dimensionless L^{2m}-norms of vorticity Ω_m $(1 \leq m < \infty)$

$$\Omega_m(t) = \left(L^{-3} \int_{\mathcal{V}} |\boldsymbol{\omega}|^{2m} \, dV \right)^{1/2m}, \qquad (4.1)$$

were then scaled into new variables D_m defined as

$$D_m(t) = \left(\varpi_0^{-1} \Omega_m \right)^{\alpha_m}, \qquad \alpha_m = \frac{2m}{4m-3}. \qquad (4.2)$$

The frequency $\varpi_0 = \nu L^{-2}$ is that of the periodic box $\mathcal{V} = [0, L]^3$. The relation between these and the first in this hierarchy, D_1, the global enstrophy, was investigated numerically by Donzis et al. (2013) and Gibbon, Donzis, Gupta, Kerr, Pandit, & Vincenzi (2014) who interpreted this relation in terms of depletion of vortex stretching. The task of this paper is to show that there is a modal dependency argument of the form

$$\Omega_m(t) = F_m \left(\Omega_1(t), \Omega_2(t), \dots, \Omega_n(t) \right) \qquad n < m, \qquad (4.3)$$

for some F_m to be determined, which is consistent with the depletion results observed and analyzed in these two papers (see section 4.3). These results show that nonlinear depletion is sufficiently strong such that solutions evolving from a wide set of initial conditions remain regular through control over D_1: note that $D_1 = \left(\varpi_0^{-1} \Omega_1 \right)^2$, the global enstrophy, is proportional to the H_1-norm of the velocity field. It is a well-known and long-standing result of Navier–Stokes analysis that in three dimensions solutions are controlled by the H_1-norm. These observations, when used in the analysis, lead to various useful estimates, particularly an equivalent energy spectrum, which turns out to be

$$\mathcal{E}(k) \sim k^{-q_{m,\lambda}} \quad \text{and} \quad q_{m,\lambda} = 3 - 4/3\lambda, \tag{4.4}$$

with a cut-off at $Lk_c = Re^{3\lambda/4}$: this spectrum is observed to decay back to $-5/3$ for large times. The numerical experiments reported in Gibbon et al. (2014) have λ lying in the range $1.15 \leq \lambda \leq 1.45$.

4.2 Definitions and a summary of numerical results

On a periodic cubical domain $[0, \, L]^3$ the incompressible, forced Navier–Stokes equations are

$$\partial_t u + u \cdot \nabla u = \nu \Delta u - \nabla p + f(\mathbf{x}), \tag{4.5}$$

together with $\operatorname{div} u = \operatorname{div} f = 0$. In terms of vorticity $\omega = \operatorname{curl} u$ these become

$$\partial_t \omega + u \cdot \nabla \omega = \nu \Delta \omega + \omega \cdot \nabla u + \operatorname{curl} f(\mathbf{x}). \tag{4.6}$$

The forcing function is taken to be L^2-bounded. The technical issue revolves around the strength of the so-called 'vortex stretching' term $\omega \cdot \nabla u$ versus the Laplacian dissipation $\Delta \omega$. The methods that have been tried over the past generation have, in effect, not been able to prove that estimates of the vortex stretching are controlled by those of the dissipation. This raises the question of what quantity can be used to measure the relative strength of this stretching. This has been achieved by considering the D_m defined above in (4.2) (Gibbon, 2010, 2011, 2012, 2013 and Donzis et al., 2013). In (4.11), the Reynolds and Grashof numbers Re and Gr are defined by

$$Gr = \frac{L^3 f_{rms}}{\nu^2}, \qquad f_{rms}^2 = L^{-3} \|f\|_2^2, \tag{4.7}$$

$$Re = \frac{L U_0}{\nu}, \qquad U_0^2 = L^{-3} \left\langle \|u\|_2^2 \right\rangle_T, \tag{4.8}$$

and the time average to time T is given by

$$\langle g(\cdot) \rangle_T = \frac{1}{T} \int_0^T g(\tau) \, d\tau. \tag{4.9}$$

Doering & Foias (2002) have introduced a simplified form of forcing with the mild restriction that involves it peaking around a length scale

ℓ, which, for simplicity, is taken here to be the box length L. Then they have shown that Navier–Stokes solutions obey $Gr \leq c\,Re^2$ and that the global enstrophy satisfies

$$\langle D_1 \rangle_T \leq GrRe + O\left(T^{-1}\right) \leq cRe^3 + O\left(T^{-1}\right). \qquad (4.10)$$

It is on this result that the proof of the following theorem is based. The main feature in the definition of the D_m is the exponent α_m defined in (4.2). Its origin lies in the fact the following result proved in (Gibbon, 2011):

Theorem 4.2.1. *For* $1 \leq m \leq \infty$

$$\langle D_m \rangle_T \leq cRe^3 + O\left(T^{-1}\right), \qquad (4.11)$$

where c is a uniform constant.

Given this result, an interesting computational question is how plots of the D_m behave against time. There are three main options:

(i) The D_m versus time curves lie on an descending scale with m;

(ii) The D_m versus time curves lie on an ascending scale with m;

(iii) For different values of m, these curves intersect.

The numerical results in Donzis et al. (2013) have suggested that the first item in the list above turns out to be result, although it must be acknowledged that this has not been proved or observed universally for all values Re and initial conditions. Moreover, the descending scale suggest that the D_m for $m > 1$ be measured against D_1. In Donzis et al. (2013) it was observed that[1],

$$D_m \leq D_1^{A_m(t)}. \qquad (4.12)$$

where $A_m(t) < \frac{1}{2}$. Numerically it was found in Gibbon et al. (2014) that a good approximation to $\max_t A_m(t)$ is

$$A_{m,\lambda} = \max_t A_m(t) = \frac{m\lambda + 1 - \lambda}{4m - 3}, \qquad (4.13)$$

where λ is a parameter that satisfies $1 \leq \lambda \leq 4$. Note that when $\lambda = 4$ then the right hand side of (4.13) is unity. This result can be interpreted as a depletion of nonlinearity. How is this relevant to the regularity problem is shown in section 4.4.

---

[1]In Donzis et al. (2013) a_m was used for the exponent in (4.12). However, a_m is easily confused with the exponent α_m in (4.2) so in Gibbon et al. (2014) this was changed to A_m.

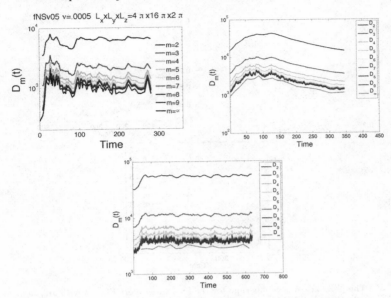

Fig. 4.1. Three log-plots of the D_m versus time with curves running from $m = 2$ & $m = 9$. D_1 is not shown. The figure in the top left position is a $1024 \times 2048 \times 512$ pseudo-spectral simulation on a long $4\pi \times 16\pi \times 2\pi$ domain with anti-parallel initial conditions (Kerr 2012, 2013, Donzis et al., 2013). The two other figures are two pseudo-spectral simulations 512^3, the first is a decaying and the second a forced NS simulation.

Figure 4.1 shows plots of $D_m(t)$ versus time (with D_1 off-scale) while Figure 4.2 shows $A_m(t) = \ln D_m / \ln D_1$ with $A_m < \frac{1}{2}$ in all cases. The numerical experiments reported in Gibbon et al. (2014) shows values of λ lying in the range $1.15 \leq \lambda \leq 1.5$.

4.3 Modal dependency and a scaling argument for $A_{m,\lambda}$

To address the problem of modal dependency a set of variables is required that can be interpreted as a set of modes or, in this case, as frequencies. The set of frequencies central to our discussion are chosen to be $\{\Omega_m(t)\}$ already defined in (4.1). In terms of the high mode/low mode idea, it could be naively speculated that a simple relationship might exist of the form expressed in (4.3). It is well known that Ω_1, the H_1-norm of the velocity field, controls the behaviour Navier–Stokes solutions so we simplify (4.3) by reducing the dependency of all Ω_m for $m \geq 2$ to just Ω_1

$$\Omega_m(t) = F_m(\Omega_1). \tag{4.14}$$

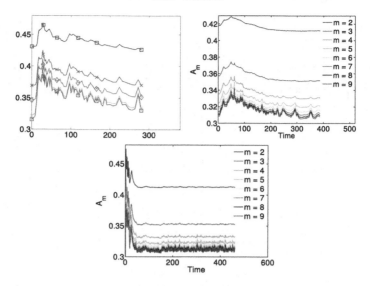

Fig. 4.2. The plots are A_m versus time for the first three plots Fig 4.1 with A_m defined in (4.13). Note that $A_m < \frac{1}{2}$.

Moreover, (4.14) is constrained and shaped by two facts, namely: (i) The Ω_m are ordered such that $\Omega_1 \leq \Omega_m$; (ii) Hölder's inequality demands that

$$1 \leq \left(\frac{\Omega_m}{\Omega_1}\right)^{m^2} \leq \left(\frac{\Omega_{m+1}}{\Omega_1}\right)^{m^2-1}. \tag{4.15}$$

These inequalities hold provided there is a solution. Keeping in mind that $\Omega_m \geq \Omega_1$, we re-write (4.14) as

$$\Omega_m = \Omega_1 f_m(\Omega_1), \tag{4.16}$$

with $f_m \geq 1$. (4.15) implies that

$$f_{m+1} \geq f_m^{\frac{m^2}{m^2-1}}. \tag{4.17}$$

We consider a class of solutions of the form

$$f_m = c_m \Omega_1^{p_m} \tag{4.18}$$

where[1]

$$p_m = (\lambda - 1)\left(\frac{m-1}{m}\right), \tag{4.19}$$

[1]The constant in the solution of (4.17) is taken to be $\lambda - 1$ to match the results in Gibbon et al. (2014).

where the c_m are chosen such that

$$c_{m+1} \geq c_m^{\left(\frac{m}{m-1}\right)^2}, \tag{4.20}$$

with $\lambda \geq 1$ as a constant parameter. Thus we have

$$\Omega_m = c_m \, \Omega_1^{1+(\lambda-1)\left(\frac{m-1}{m}\right)}, \tag{4.21}$$

which, when re-written in the D_m-format, is

$$D_m = c_m D_1^{\left(\frac{m+(\lambda-1)(m-1)}{4m-3}\right)} \equiv c_m D_1^{A_{m,\lambda}}, \tag{4.22}$$

$$A_{m,\lambda} = \frac{m\lambda - \lambda + 1}{4m-3}. \tag{4.23}$$

This result is merely a scaling argument based on the assumption made in (4.16). Nevertheless, while $A_{m,\lambda}$ in (4.22) is consistent with (4.13), the direction of the inequality (4.12) lies with D_m on the left. This suggests that the maxima in time of the numerical D_m-curves may wish to saturate the inequalities in (4.15) although there could be considerable slack due to the multiplicative constants c_m which increase with m according to (4.20).

4.4 Regularity in the $D_1 - D_m$ plane

To estimate how the depletion of nonlinearity expressed in (4.13) works we consider the vortex stretching term that occurs in the differential inequality[1] for D_1

$$D_1 = \varpi_0^{-2} L^{-3} \int_{\mathcal{V}} |\boldsymbol{\omega}|^2 dV. \tag{4.24}$$

Thus we have

$$\frac{1}{2} \frac{d}{dt} D_1 = \varpi_0^{-2} L^{-3} \left\{ -\nu \int_{\mathcal{V}} |\nabla \boldsymbol{\omega}|^2 \, dV + \int_{\mathcal{V}} |\nabla \boldsymbol{u}||\boldsymbol{\omega}|^2 \, dV + L^{-1} \|\boldsymbol{f}\|_2 \|\boldsymbol{\omega}\|_2 \right\}. \tag{4.25}$$

To estimate the vortex stretching (second) term we write

$$\int_{\mathcal{V}} |\nabla \boldsymbol{u}||\boldsymbol{\omega}|^2 \, dV = \int_{\mathcal{V}} |\boldsymbol{\omega}|^{\frac{2m-3}{m-1}} |\boldsymbol{\omega}|^{\frac{1}{m-1}} |\nabla \boldsymbol{u}| \, dV$$

$$\leq \left(\int_{\mathcal{V}} |\boldsymbol{\omega}|^2 \, dV \right)^{\frac{2m-3}{2(m-1)}} \left(\int_{\mathcal{V}} |\boldsymbol{\omega}|^{2m} \, dV \right)^{\frac{1}{2m(m-1)}} \left(\int_{\mathcal{V}} |\nabla \boldsymbol{u}|^{2m} \, dV \right)^{\frac{1}{2m}}.$$

---

[1]This uses the contradiction method explained in Donzis et al. (2014).

Therefore

$$\int_{\mathcal{V}} |\nabla \boldsymbol{u}||\boldsymbol{\omega}|^2 \, \mathrm{d}V \leq c_m \left(\int_{\mathcal{V}} |\boldsymbol{\omega}|^2 \, \mathrm{d}V \right)^{\frac{2m-3}{2(m-1)}} \left(\int_{\mathcal{V}} |\boldsymbol{\omega}|^{2m} \, \mathrm{d}V \right)^{\frac{1}{2(m-1)}}$$

$$= c_m L^3 \varpi_0^3 D_1^{\frac{2m-3}{2m-2}} D_m^{\frac{4m-3}{2m-2}}, \tag{4.26}$$

based on $\|\nabla \boldsymbol{u}\|_p \leq c_p \|\boldsymbol{\omega}\|_p$, for $1 < p < \infty$. Therefore, using the depletion $D_m \leq D_1^{A_{m,\lambda}}$,

$$L\nu^{-2} \int_{\mathcal{V}} |\nabla \boldsymbol{u}||\boldsymbol{\omega}|^2 \, \mathrm{d}V \leq c_m \varpi_0 D_1^{\xi_{m,\lambda}}, \tag{4.27}$$

where $\xi_{m,\lambda}$ is given by

$$\xi_{m,\lambda} = \frac{m\lambda + 1 - \lambda + 2m - 3}{2(m-1)} = 1 + \tfrac{1}{2}\lambda. \tag{4.28}$$

Moreover, integration by parts and a Schwarz inequality give

$$\int_{\mathcal{V}} |\nabla \boldsymbol{\omega}|^2 \, \mathrm{d}V \geq \left(\int_{\mathcal{V}} |\boldsymbol{\omega}|^2 \, \mathrm{d}V \right)^2 \left(\int_{\mathcal{V}} |\boldsymbol{u}|^2 \, \mathrm{d}V \right)^{-1}, \tag{4.29}$$

and so we end up with

$$\frac{1}{2} \frac{\mathrm{d}}{\mathrm{d}t} D_1 \leq \varpi_0 \left(-\frac{D_1^2}{E} + c_m D_1^{1+\frac{1}{2}\lambda} + Gr D_1^{1/2} \right), \tag{4.30}$$

where the dimensionless energy E is defined by

$$E = L^{-1}\nu^{-2} \int_{\mathcal{V}} |\boldsymbol{u}|^2 \, \mathrm{d}V \leq Gr^2, \tag{4.31}$$

and has an upper bound $\overline{\lim}_{t\to\infty} E \leq Gr^2$ which is easily found by standard methods. With this we have:

Proposition 4.4.1. *(Gibbon et al., 2014): If the solution always remains in the region $1 \leq \lambda < 2$, there exists an absorbing ball for D_1 of radius*

$$\overline{\lim}_{t\to\infty} D_1 \leq c_m Gr^{\frac{4}{2-\lambda}} + O\left(Gr^{4/3}\right). \tag{4.32}$$

The range of control over D_1 in $1 \leq \lambda < 2$ can be extended to $\lambda = 2$ as (4.30) shows that there is an exponentially growing bound on D_1 at this value.

Coming from a different direction, Lu & Doering (2008) used a numerical calculus of variations argument to find the value(s) of the exponent $\xi_{m,\lambda}$ when the rate of enstrophy production is maximized subject to the

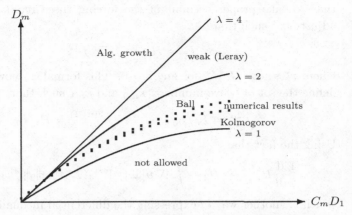

Fig. 4.3. The $D_1 - D_m$ plane represented by the inequalities in (4.12) at some value of $m > 1$ illustrating the concave tongues parametrised by λ. $\lambda = 1$ is the extreme lower bound just above which lie solutions corresponding to Kolmogorov's statistical theory. An absorbing ball for D_1 exists in the range $1 \leq \lambda < 2$ while Leray's weak solutions exist in $2 \leq \lambda < 4$. The region beyond the line $D_m = D_1$ has been shown in Gibbon et al. (2014) has also been shown to be regular but may correspond to pathological initial data.

constraint div $\boldsymbol{u} = 0$. They found that two branches existed, the lower being $D_1^{1.78}$ and the uppermost $D_1^{2.997}$. Later, Schumacher, Eckhardt, & Doering (2010) suggested that $7/4$ and 3 were the likely values of these two exponents; the exponent $\xi_{m,\lambda} = 7/4$ corresponds to $\lambda = 1.5$ which lies at the upper end of our observed range $1.15 \leq \lambda \leq 1.5$.

4.5 Energy spectra and intermittency

The results summarized in the previous section and expressed in Fig. 4.3 raise the question how close or far these results are from those that arise in Kolmogorov's theory of turbulence (Frisch, 1995)? One way of testing this is to look at estimates for the energy spectrum. Kolmogorov's theory is based on statistical averages whereas the Navier–Stokes equations are PDEs. Despite this obstacle, Doering & Gibbon (2002) have shown how to associate bounds of time averages with the moments of an energy spectrum $\mathcal{E}(k)$. The following results can be found in Gibbon et al. (2014) except we now concentrate on the region $1 \leq \lambda \leq 2$ in Fig. 4.3.

In the standard manner (Doering & Gibbon, 1995, 2002), we define

$$H_n(t) = \int_{\mathcal{V}} |\nabla^n \boldsymbol{u}|^2 \, dV \qquad \text{with} \qquad H_0 = \int_{\mathcal{V}} |\boldsymbol{u}|^2 \, dV, \qquad (4.33)$$

where the label n refers to derivatives. Doering & Gibbon (2002) showed that to take proper account of the forcing these require an additive adjustment such that

$$F_n = H_n + \tau^2 \|\nabla^n \boldsymbol{f}\|_2^2, \qquad (4.34)$$

where $\tau^{-1} \sim \varpi_0 Gr^{\frac{1}{2}+\varepsilon}$ for any $\varepsilon > 0$. This formalism now allows us to define the set of 'wave-numbers' $\kappa_{n,0}$ and $\kappa_{n,1}$ such that

$$\kappa_{n,0}^{2n} = F_n/F_0, \qquad \kappa_{n,1}^{2(n-1)} = F_n/F_1. \qquad (4.35)$$

Using the fact that

$$\frac{1}{2}\frac{d}{dt}H_1 \leq -\nu H_2 + \int_{\mathcal{V}} |\nabla \boldsymbol{u}||\boldsymbol{\omega}|^2 \, dV + L^{-1}\|\boldsymbol{\omega}\|_2\|\boldsymbol{f}\|_2, \qquad (4.36)$$

which is another way of expressing the differential inequality for D_1, we can re-visit the inequality in (4.26) to estimate the integral in (4.36) with the application of the depletion expressed in (4.12)

$$\int_{\mathcal{V}} |\nabla \boldsymbol{u}||\boldsymbol{\omega}|^2 \, dV \leq \varpi_0 \left(L^3 \varpi_0^2\right)^{1-\xi_{m,\lambda}} H_1^{\xi_{m,\lambda}} = \varpi_0 H_1 D_1^{\xi_{m,\lambda}-1}, \quad (4.37)$$

which, again, is just another expression of (4.27). The bounds $1 \leq \lambda < 2$ mean that

$$3/2 \leq \xi_{m,\lambda} < 2, \qquad (4.38)$$

and so

$$\frac{1}{2}\frac{d}{dt}H_1 \leq \varpi_0 \left\{-L^2 H_2 + H_1 D_1^{\xi_{m,\lambda}-1}\right\} + L^{-1}\|\boldsymbol{\omega}\|_2\|\boldsymbol{f}\|_2, \qquad (4.39)$$

which, when the H_n are adjusted to the F_n defined in (4.34) as in Doering & Gibbon (2002), becomes

$$\frac{1}{2}\frac{d}{dt}F_1 \leq \varpi_0 \left\{-L^2 F_2 + F_1 D_1^{\xi_{m,\lambda}-1}\right\} + c_n \varpi_0 Gr\, F_1. \qquad (4.40)$$

Dividing (4.40) by F_1 and time averaging gives

$$L^2 \langle \kappa_{2,1}^2 \rangle_T \leq \langle D_1 \rangle_T^{\xi_{m,\lambda}-1} \leq c\, Re^{3(\xi_{m,\lambda}-1)}. \qquad (4.41)$$

Moreover, we can also write

$$\langle \kappa_{2,0} \rangle_T \leq \langle \kappa_{2,1}\kappa_{1,0} \rangle_T^{1/2} \leq \langle \kappa_{2,1}^2 \rangle_T^{1/4} \langle \kappa_{1,0}^2 \rangle_T^{1/4}. \qquad (4.42)$$

In Doering & Gibbon (2002) it was shown that Leray's energy inequality leads to an estimate for $L^2 \langle \kappa_{1,0}^2 \rangle_T \leq Re^{1+\varepsilon}$ (we ignore the infinitesimal $\varepsilon > 0$). This is combined with (4.42) to show that

$$\langle \kappa_{2,0} \rangle_T \leq c\, Re^{\sigma_{m,\lambda}} + O\left(Gr^{1/4}\right), \qquad (4.43)$$

where

$$\sigma_{m,\lambda} = \frac{3(\chi_{m,\lambda} - 1) + 2(m-1)}{8(m-1)} = (3\lambda + 2)/8. \qquad (4.44)$$

To interpret this in terms of statistical turbulence theory (restricting attention to forcing at the longest wavelength $\ell = L$), suppose that Gr is high enough and the resulting flow is turbulent, ergodic and isotropic enough in the limit $T \to \infty$ that the wave-numbers $\langle \kappa_{n,0} \rangle_T$ may be identified with the moments of the energy spectrum $\mathcal{E}(k)$ according to

$$\langle \kappa_{n,0} \rangle_T := \left(\frac{\int_{L^{-1}}^{\infty} k^{2n} \mathcal{E}(k) \, \mathrm{d}k}{\int_{L^{-1}}^{\infty} \mathcal{E}(k) \, \mathrm{d}k} \right)^{1/2n}. \qquad (4.45)$$

The *a priori* constraints on $\mathcal{E}(k)$ are that the velocity U and energy dissipation rate ϵ obey

$$U^2 = \int_{L^{-1}}^{\infty} \mathcal{E}(k) \, \mathrm{d}k, \qquad \epsilon = \int_{L^{-1}}^{\infty} \nu k^2 \mathcal{E}(k) \, \mathrm{d}k. \qquad (4.46)$$

Suppose also that $\mathcal{E}(k)$ displays an "inertial range" in the sense that it scales with a power of k up to an effective cut-off wavenumber k_c. For simplicity, let us write

$$\mathcal{E}(k) = \begin{cases} A k^{-q}, & L^{-1} \leq k \leq k_c, \\ 0, & k > k_c. \end{cases} \qquad (4.47)$$

We also assume that k_c diverges as $\nu \to 0$, while U^2 and ϵ remain finite, and that A depends only upon the energy flux ϵ and the outer length scale $\ell = L$. Then we have the asymptotic relations

$$\epsilon \sim \frac{U^3}{L} \qquad \text{and} \qquad L k_c \sim \left(\frac{\epsilon}{\nu^3} \right)^{\frac{1}{9-3q}} L^{\frac{4}{9-3q}} \sim Re^{\frac{1}{3-q}}. \qquad (4.48)$$

Then the moments of the spectrum $\langle \kappa_{n,0} \rangle_T$ satisfy

$$L \langle \kappa_{n,0} \rangle_T \sim (L k_c)^{1 - \frac{q-1}{2n}} \sim Re^{\frac{1}{3-q} - \frac{1}{2n}\left(\frac{q-1}{3-q}\right)}. \qquad (4.49)$$

Now let us compare this scaling result with the estimate in (4.43) for $n = 2$ with $q = q_{m,\lambda}$: this correspondence tells us that

$$q_{m,\lambda} = \frac{12\sigma_{m,\lambda} - 5}{4\sigma_{m,\lambda} - 1} = 3 - \frac{4}{3\lambda}. \qquad (4.50)$$

Thus in the three cases $\lambda = 1$, $\lambda = 2$ and $\lambda = 4$ we have

$$q_{m,1} = 5/3, \qquad q_{m,2} = 7/3, \qquad q_{m,4} = 8/3. \qquad (4.51)$$

The $5/3$ at the lower end is the conventional Kolmogorov result which rises to just under $7/3$ at the end of the regular region in Fig. 4.3. The cut-off of the inertial range as (4.47) is given by

$$Lk_c = Re^{3\lambda/4}, \qquad (4.52)$$

which corresponds to

$$Lk_{c,1} = Re^{3/4}, \qquad Lk_{c,2} = Re^{3/2}, \qquad Lk_{c,4} = Re^3. \quad (4.53)$$

It is clear therefore that results equivalent to Kolmogorov's theory (see Frisch, 1995, Boffetta, Mazzino, & Vulpiani, 2008, Pandit, Perlekar, & Ray, 2009) lie right near the lower boundary at $\lambda = 1$ and, as λ increases away from unity, both the spectrum and the cut-off depart from this according to (4.50) and (4.52), with corresponding solutions becoming more intermittent as λ increases from unity. However, these formulae have been estimated using the form of $A_{m,\lambda}$ that is a fit to the maxima of the D_m-plots. Once decay sets in for the unforced cases, q_m settles back to close to $-5/3$, as is usually observed in simulations. Beyond $\lambda = 2$ lies the region where only Leray's weak solutions have been proven to exist.

In Doering & Gibbon (2002) it was shown that for $n \geq 2$,

$$\langle \kappa_{n,0} \rangle_T \leq c\, Re^{3 - \frac{5}{2n}}. \qquad (4.54)$$

For $n = 2$ this means $\sigma_{m,\lambda} = 7/4$ and thus $q_{m,\lambda} = 8/3$ which is the same as (4.52) at $\lambda = 4$. Sulem & Frisch (1975) showed that a $k^{-8/3}$ energy spectrum is the borderline steepness capable of sustaining an energy cascade. This spectrum corresponds to the extreme limit, where the energy dissipation is concentrated on sets of dimension zero (points) in space.

4.6 An estimate for the attractor dimension in the depleted regime

The existence of an H_1-ball of finite radius has been demonstrated in Proposition 4.4.1, subject to the depletion in regime I. From this follows the existence of a global attractor $\mathcal{A}$ in this regime. In the following two subsections it will be shown how this leads to an estimate for the Lyapunov dimension of $\mathcal{A}$ written as $d_L(\mathcal{A})$. Such estimates are useful as they connect the number of degrees of freedom $\mathcal{N}_d$ in d spatial dimensions with the resolution length ℓ_d

$$\left(\frac{L}{\ell_d}\right)^d = \mathcal{N}_d. \tag{4.55}$$

Essentially the left hand side of (4.55) is the number of small 'vortical features' of volume ℓ_d^d in the system volume L^d. $\mathcal{N}_d$ can be interpreted many ways: one way is to take $\mathcal{N}_d$ as the global attractor dimension $d_L(\mathcal{A})$ while another is to take it as the number of determining modes or nodes as in Jones & Titi (1993). A further independent way of estimating ℓ_3, specifically for the 3D Navier–Stokes equations, is to take

$$L\ell_3^{-1} \equiv \langle \kappa_{2,0} \rangle \le c\, Re^{\sigma_{m,\lambda}} \tag{4.56}$$

as in (4.43), where $\sigma_{m,\lambda} = (3\lambda + 2)/8$.

4.6.1 Global Lyapunov exponents and the Kaplan–Yorke formula

The idea of the Lyapunov exponents of a nonlinear system connect the system dynamics with the attractor dimension through what is known as the Kaplan–Yorke formula. For systems of ODEs the Lyapunov exponents control the exponential growth or contraction of volume elements in phase space. The Kaplan–Yorke formula then expresses the balance between volume growth and contraction realized on the attractor. In the case of PDEs the concept has been rigorously applied to global attractors by Constantin & Foias (1985, 1988): see also Temam (1988), Doering & Gibbon (1995), Gibbon & Titi (1997). For Lyapunov exponents labelled in descending order and designated by μ_n, the Lyapunov dimension d_L is defined by

$$d_L = N - 1 + \frac{\mu_1 + \ldots + \mu_{N-1}}{-\mu_N}, \tag{4.57}$$

where the number N of μ_n is chosen to satisfy

$$\sum_{n=1}^{N-1} \mu_n \ge 0 \qquad \text{but} \qquad \sum_{n=1}^{N} \mu_n < 0. \tag{4.58}$$

Note that according to the definition of N, the ratio of exponents in (4.57) satisfies

$$0 \le \frac{\mu_1 + \ldots + \mu_{N-1}}{-\mu_N} < 1. \tag{4.59}$$

Clearly d_L is bounded above and below by $N - 1 \leq d_L < N$. Thus the value of N that turns the sign of the sum of the Lyapunov exponents, as in (4.58), is that value of N that bounds above d_L. To use the method for PDEs as developed in Constantin & Foias (1985, 1988) the phase space is replaced by $\boldsymbol{u} \in L^2 \cap \operatorname{div} \boldsymbol{u} = 0$, which is infinite dimensional. The velocity field $\boldsymbol{u}(t)$ forms an orbit in this space, with different sets of initial conditions $\boldsymbol{u}(0) + \delta \boldsymbol{u}_i(0)$, which evolve into $\boldsymbol{u}(t) + \delta \mathbf{u}_i(t)$ for $i = 1, \ldots, N$. The linearized form of the Navier–Stokes equations in terms of $\delta \boldsymbol{u}$ of $\boldsymbol{u}$ is

$$\partial_t(\delta \boldsymbol{u}) + \boldsymbol{u} \cdot \nabla \delta \boldsymbol{u} + \delta \boldsymbol{u} \cdot \nabla \boldsymbol{u} = \nu \Delta \delta \boldsymbol{u} - \nabla \delta p \,, \qquad (4.60)$$

which can also be written in the form

$$\partial_t(\delta \boldsymbol{u}) = \mathcal{M} \delta \boldsymbol{u} \,. \qquad (4.61)$$

If they are chosen to be linearly independent, initially these $\delta \boldsymbol{u}_i$ form an N-volume or parallelpiped of volume

$$V_N(t) = |\delta \boldsymbol{u}_1 \wedge \delta \boldsymbol{u}_2 \ldots \wedge \delta \boldsymbol{u}_N| \,. \qquad (4.62)$$

It is now necessary to find the time evolution of V_N. This is given by

$$\frac{\mathrm{d}}{\mathrm{d}t} V_N = V_N \operatorname{Tr} \left[\mathbf{P}_N \mathcal{M} \mathbf{P}_N \right] \,, \qquad (4.63)$$

which is easily solved to give

$$V_N(t) = V_N(0) \exp \left[\int_0^t \operatorname{Tr} \left[\mathbf{P}_N \mathcal{M} \mathbf{P}_N \right] (\tau) \, \mathrm{d}\tau \right] \,. \qquad (4.64)$$

$\mathbf{P}_N(t)$ is an L^2-orthogonal projection, using the orthonormal set of functions $\{\boldsymbol{\phi}_i\}$, onto the finite dimensional subspace $\mathbf{P}_N L^2$, which spans the set of vectors $\delta \boldsymbol{u}_i$ for $i = 1, \ldots, N$. In terms of the time average $\langle \cdot \rangle_t$ up to time t, the sum of the first N global Lyapunov exponents is taken to be (Constantin & Foias, 1985, 1988)

$$\sum_{n=1}^{N} \mu_n = \langle \operatorname{Tr} \left[\mathbf{P}_N \mathcal{M} \mathbf{P}_N \right] \rangle_t \,. \qquad (4.65)$$

4.6.2 Estimating $d_L(\mathcal{A})$

In this subsection we follow the calculation Gibbon & Titi (1997) and Gibbon et al. (2014). We wish to find the value of N that turns the sign of $\langle \mathrm{Tr}\,[\mathbf{P}_N \mathcal{M} \mathbf{P}_N] \rangle_t$ and for which volume elements contract to zero. This value of N bounds above d_L as in (4.57). To estimate this we write

$$\mathrm{Tr}\,[\mathbf{P}_N \mathcal{M} \mathbf{P}_N] = \sum_{n=1}^{N} \int_{\mathcal{V}} \boldsymbol{\phi}_n \cdot \{\nu \Delta \boldsymbol{\phi}_n - \boldsymbol{u} \cdot \nabla \boldsymbol{\phi}_n - \boldsymbol{\phi}_n \cdot \nabla \boldsymbol{u} - \nabla \widetilde{p}(\boldsymbol{\phi}_n)\} \, \mathrm{d}V.$$
(4.66)

Since $\mathrm{div}\,\delta_m \boldsymbol{u}_n = 0$ for all n, then $\mathrm{div}\,\boldsymbol{\phi}_n = 0$ also and so the pressure term integrates away, as does the second term

$$\mathrm{Tr}\,[\mathbf{P}_N \mathcal{M} \mathbf{P}_N] \leq -\nu \sum_{n=1}^{N} \int_{\mathcal{V}} |\nabla \boldsymbol{\phi}_n|^2 \, \mathrm{d}V + \sum_{n=1}^{N} \int_{\mathcal{V}} |\nabla \boldsymbol{u}|\, |\boldsymbol{\phi}_n|^2 \, \mathrm{d}V. \quad (4.67)$$

Because the $\boldsymbol{\phi}_n$ are orthonormal they obey the relations

$$\sum_{n=1}^{N} \int_{\mathcal{V}} |\boldsymbol{\phi}_n|^2 \, \mathrm{d}V = N, \quad \text{and} \quad \mathrm{Tr}\,[\mathbf{P}_N(-\Delta)\mathbf{P}_N] = \sum_{n=1}^{N} \int_{\mathcal{V}} |\nabla \boldsymbol{\phi}_n|^2 \, \mathrm{d}V.$$
(4.68)

In $3D$ the $\boldsymbol{\phi}_n$ satisfy the Lieb–Thirring inequalities for orthonormal functions – see Constantin & Foias (1985, 1988), Temam (1988), Foias et al. (2001).

$$\int_{\mathcal{V}} \left(\sum_{n=1}^{N} |\boldsymbol{\phi}_n|^2 \right)^{5/3} \mathrm{d}V \leq c \sum_{n=1}^{N} \int_{\mathcal{V}} |\nabla \boldsymbol{\phi}_n|^2 \, \mathrm{d}V, \quad (4.69)$$

where c is independent of N. Moreover, it is known that the first N eigenvalues of the Stokes operator in three-dimensions satisfy

$$\mathrm{Tr}\,[\mathbf{P}_N(-\Delta)\mathbf{P}_N] \geq c\, N^{5/3} L^{-2}. \quad (4.70)$$

To exploit the Lieb–Thirring inequality (4.69) to estimate the last term in (4.67) we write it as

$$\sum_{n=1}^{N} \int_{\mathcal{V}} |\nabla \boldsymbol{u}|\, |\boldsymbol{\phi}_n|^2 \, \mathrm{d}V \leq \left[\int_{\mathcal{V}} |\nabla \boldsymbol{u}|^{5/2} \, \mathrm{d}V \right]^{2/5} \left[\int_{\mathcal{V}} \left(\sum_{n=1}^{N} |\boldsymbol{\phi}_n|^2 \right)^{5/3} \mathrm{d}V \right]^{3/5}.$$
(4.71)

Hence, using (4.69) and time averaging $\langle \cdot \rangle_t$, we find

$$\left\langle \sum_{n=1}^{N} \int_{\mathcal{V}} |\nabla \boldsymbol{u}|\, |\boldsymbol{\phi}_n|^2 \, \mathrm{d}V \right\rangle_t \quad (4.72)$$

$$\leq c \left\langle (\mathrm{Tr}\,[\mathbf{P}_N(-\Delta)\mathbf{P}_N])^{3/5} \left(\int_{\mathcal{V}} |\nabla \boldsymbol{u}|^{5/2} \, \mathrm{d}V \right)^{2/5} \right\rangle_t$$

$$\leq \tfrac{3\nu}{5} \langle \mathrm{Tr}\,[\mathbf{P}_N(-\Delta)\mathbf{P}_N] \rangle_t + \tfrac{2c}{5\nu^{3/2}} \left\langle \int_{\mathcal{V}} |\nabla \boldsymbol{u}|^{5/2} \, \mathrm{d}V \right\rangle_t \quad (4.73)$$

and so (4.67) can be written as

$$\langle \text{Tr}\,[\mathbf{P}_N \mathcal{M} \mathbf{P}_N] \rangle_t \leq -\frac{2}{5}\nu\,\langle \text{Tr}\,[\mathbf{P}_N(-\Delta)\mathbf{P}_N] \rangle_t + \frac{2}{5}c\,\nu^{-3/2}\left\langle \int_{\mathcal{V}} |\nabla u|^{5/2}\,dV \right\rangle_t .$$
(4.74)

To estimate the nonlinear term we use Hölder's inequality to obtain $(m > 1)$

$$\begin{aligned}
\int_{\mathcal{V}} |\nabla u|^{5/2}\,dV &\leq c\int_{\mathcal{V}} |\boldsymbol{\omega}|^{5/2}\,dV \\
&\leq c\left(\int_{\mathcal{V}} |\boldsymbol{\omega}|^2\,dV\right)^{\frac{4m-5}{4(m-1)}}\left(\int_{\mathcal{V}} |\boldsymbol{\omega}|^{2m}\,dV\right)^{\frac{1}{4(m-1)}} \\
&\leq c\,\varpi_0^{5/2} L^3 D_1^{\frac{4m-5}{4(m-1)}} D_m^{\frac{4m-3}{4(m-1)}} .
\end{aligned}$$
(4.75)

Therefore, using this and (4.70), we find

$$\varpi_0^{-1}\langle \text{Tr}\,[\mathbf{P}_N \mathcal{M} \mathbf{P}_N] \rangle_t \leq -c_1 N^{5/3} + c_2\left\langle D_1^{\frac{4m-5}{4(m-1)}} D_m^{\frac{4m-3}{4(m-1)}} \right\rangle_t .$$
(4.76)

The depletion of nonlinearity $D_m \leq D_1^{A_{m,\lambda}}$ with

$$\chi_{m,\lambda} = A_{m,\lambda}(4m-3)$$
(4.77)

is used again in (4.75), thereby giving

$$\left\langle D_1^{\frac{4m-5}{4(m-1)}} D_m^{\frac{4m-3}{4(m-1)}} \right\rangle_t \leq \langle D_1 \rangle_t\,(\overline{\lim}_{t\to\infty} D_1)^{\frac{\chi_{m,\lambda}-1}{4(m-1)}} ,$$
(4.78)

where $\chi_{m,\lambda} = A_{m,\lambda}(4m-3)$. Proposition 4.4.1 and the estimate $\langle D_1 \rangle_t \leq c\,Gr\,Re$ (Doering & Foias, 2002) then allow us to write

$$\langle D_1 \rangle_t\,(\overline{\lim}_{t\to\infty} D_1)^{\frac{\chi_{m,\lambda}-1}{4(m-1)}} \leq c\,(Gr\,Re)Gr^{\frac{\chi_{m,\lambda}-1}{2m-1-\chi_{m,\lambda}}} \leq c\,Re^{\frac{6m-5-\chi_{m,\lambda}}{2m-1-\chi_{m,\lambda}}} ,$$
(4.79)

and so (4.76) can be written as

$$\langle \text{Tr}\,[\mathbf{P}_N \mathcal{M} \mathbf{P}_N] \rangle_t \leq \varpi_0\left(-c_1 N^{5/3} + c_2 Re^{\frac{6m-5-\chi_{m,\lambda}}{2m-1-\chi_{m,\lambda}}}\right).$$
(4.80)

To find an estimate solely in terms of Gr the $(Gr\,Re)$-term of (4.79) is replaced by Gr^2. Choosing $\chi_{m,\lambda}$ as in (4.77), we have proved:

Proposition 4.6.1. *If a solution always remains in regime I the Lyapunov dimension of the global attractor $\mathcal{A}$ is estimated as*

$$d_L(\mathcal{A}) \leq c_{1,m} Re^{\frac{3}{5}\left(\frac{6-\lambda}{2-\lambda}\right)} ,$$
(4.81)

or, alternatively, as

$$d_L(\mathcal{A}) \le c_{2,m} Gr^{\frac{3}{5}\left(\frac{4-\lambda}{2-\lambda}\right)}.\tag{4.82}$$

Note that when $\lambda = 1$ (see Figure 4.3) then $d_L(\mathcal{A}) \le c_{1,m} Re^3$. From (4.81) the estimate for the resolution length ℓ is given by

$$L\ell^{-1} \le c_{3,m} Re,\tag{4.83}$$

but becomes significantly worse as λ increases:

$$L\ell^{-1} \le c_{4,m} Re^{\frac{1}{5}\left(\frac{6-\lambda}{2-\lambda}\right)},\tag{4.84}$$

Note that this is significantly worse than the estimate $Re^{(3\lambda+2)/8}$ in (4.56).

4.7 Conclusion

Regimes of regularity and non-regularity has been set out pictorially as in Fig. 4.3, but this has been based on the mathematical assumption that Ω_1 controls higher Ω_m through (4.3) or the simpler form of (4.16). The parametrization of this depletion by λ comes from a scaling argument on the Hölder inequalities which comes out to be (4.21) and is repeated here

$$\Omega_m = c_m \Omega_1^{1+(\lambda-1)\left(\frac{m-1}{m}\right)}.\tag{4.85}$$

The value of λ then shows how strong this depletion has to be to gain control over D_1 from which existence and uniqueness of solutions follow provided solutions remain in the $1 \le \lambda \le 2$ region. $\lambda = 1$ is the lower bound $\Omega_1 = \Omega_m$ but, as λ increases to $\lambda < 2$, control over D_1 is retained until the critical value $\lambda = 2$ is reached: here D_1 is controlled by an exponentially growing bound in time. For the regime $2 < \lambda < 4$ control over D_1 is lost, leaving us with the fallback of Leray's weak solutions. At $\lambda = 4$, $A_m = 1$ in which case $D_m = D_1$ from which there appears to be no gain. The arguments of section 4.5 show that results for the equivalent energy spectrum and its cut-off $Lk_{c,\lambda} = Re^{3\lambda/4}$ are in accord with Kolmogorov's theory at the minimum at $\lambda = 1$ but increase as the flow becomes more intermittent until the critical value $\lambda = 2$ is reached. Since the spectral estimates are based on the peak values of $A_m(t)$ it is not surprising that the spectrum in numerical simulations decay back to $-5/3$ at longer times.

These results, as summarized in Fig. 4.3, suggest that stability is an issue: in other words, if one begins with initial data in the region $1 < \lambda < 2$ does the solution always remain there? Likewise can the same be said for the region $2 < \lambda < 4$?

Acknowledgements: I am grateful to my collaborators Diego Donzis, Anupam Gupta, Rahul Pandit, Robert Kerr and Dario Vincenzi for a selection of figures and various discussions on the issues raised in this paper.

References

Boffetta, G., Mazzino, A., & Vulpiani, A. (2008) Twenty-five years of multi-fractals in fully developed turbulence: a tribute to Giovanni Paladin. *J. Phys. A* **41**, 363001.

Constantin, P. & Foias, C. (1985) Global Lyapunov exponents, Kaplan-Yorke formulas and the dimension of the attractors for $2D$ Navier–Stokes Equations. *Comm. Pure Applied Math.* **38**, 1–27.

Constantin, P. & Foias, C. (1988) *The Navier–Stokes equations.* Chicago University Press, Chicago, USA.

Doering, C.R. (2009) The $3D$ Navier–Stokes problem. *Annu. Rev. Fluid Mech.* **41**, 109–128.

Doering, C.R. & Foias, C. (2002) Energy dissipation in body-forced turbulence. *J. Fluid Mech.* **467**, 289–306.

Doering C.R. & Gibbon, J.D. (1995) *Applied analysis of the Navier–Stokes equations.* Cambridge University Press, Cambridge, England.

Doering, C.R. & Gibbon, J.D. (2002) Bounds on moments of the energy spectrum for weak solutions of the 3D Navier–Stokes equations. *Physica D* **165**, 163–175.

Donzis, D., Yeung, P.K., & Sreenivasan, K.R. (2008) Dissipation and enstrophy in isotropic turbulence: scaling and resolution effects in direct numerical simulations. *Phys. Fluids* **20**, 045108.

Donzis, D. & Yeung, P.K. (2010) Resolution effects and scaling in numerical simulations of passive scalar mixing in turbulence. *Phys. D* **239**, 1278–1287.

Donzis, D., Gibbon, J.D., Gupta, A., Kerr, R.M., Pandit, D., & Vincenzi, D. (2013) Vorticity moments in four numerical simulations of the $3D$ Navier–Stokes equations. *J. Fluid Mech.* **732**, 316 – 331.

Escauriaza, L., Seregin, G., & Šverák, V. (2003) L^3-solutions to the Navier–Stokes equations and backward uniqueness. *Russ. Math. Surveys* **58**, 211–250.

Eswaran, V. & Pope, S.B. (1988) An examination of forcing in direct numerical simulations of turbulence. *Comput. Fluids* **16**, 257–278.

Farhat, A., Jolly, M.S., & Titi, E.S. (2015) Continuous data assimilation for $2D$ Bénard convection through velocity measurements alone. *Phys. D* **303**, 59–66.

Foias, C., Jolly, M.S., Kevrekidis, I.G., Sell, G.R., & Titi, E.S. (1988) On the computation of inertial manifolds. *Phys. Lett. A* **131**, 433–436.

Foias, C., Manley, O., Rosa, R., & Temam, R. (2001) *Navier–Stokes equations and turbulence.* Cambridge University Press, Cambridge, England.

Foias, C., Sell, G. R., & Temam, R. (1988) Inertial manifolds for nonlinear evolutionary equations. *J. Diff. Equ.* **73**, 309–353.

Foias, C. & Temam, R. (1984) Determination of the solutions of the Navier–Stokes equations by a set of nodal values. *Math Comp.* **43**, 117–133.

Foias, C. & Titi, E.S. (1991) Determining nodes, finite difference schemes and inertial manifolds. *Math Comp.* **4**, 134–153.

Frisch, U. (1995) *Turbulence: the legacy of A.N. Kolmogorov.* Cambridge University Press, Cambridge, England.

Gibbon, J.D., Donzis, D., Gupta, A., Kerr, R.M., Pandit R., & Vincenzi, D. (2014) Regimes of nonlinear depletion and regularity in the 3D Navier–Stokes equations. *Nonlinearity* **27**, 1–19.

Gibbon, J.D. (2010) Regularity and singularity in solutions of the three-dimensional Navier–Stokes equations. *Proc. Royal Soc A* **466**, 2587–2604.

Gibbon, J.D. (2011) A hierarchy of length scales for weak solutions of the three-dimensional Navier–Stokes equations. *Comm Math. Sci.* **10**, 131–136.

Gibbon, J.D. (2012) Conditional regularity of solutions of the three dimensional Navier–Stokes equations & implications for intermittency. *J. Math. Phys.* **53**, 115608.

Gibbon, J.D. (2013) Dynamics of scaled vorticity norms for the three-dimensional Navier–Stokes and Euler equations. *Procedia IUTAM* **7**, 39–48.

Gibbon, J.D. & Titi, E.S. (1997) Attractor dimension and small length scale estimates for the 3D Navier–Stokes equations, *Nonlinearity* **10**, 109–119.

Guckenheimer, J. & Holmes, P. (1997) *Nonlinear Oscillations, Dynamical Systems, and Bifurcations of Vector Fields.* Applied Mathematical Sciences **42**, Springer-Verlag, Berlin.

Ishihara, T., Gotoh, T., & Kaneda, Y. (2009) Study of high-Reynolds number isotropic turbulence by direct numerical simulation. *Annu. Rev. Fluid Mech.* **41**, 16–180.

Jimenez, J., Wray, A.A., Saffman, P.G., & Rogallo, R.S. (1993) The structure of intense vorticity in isopropic turbulence. *J. Fluid Mech.* **255**, 65–91.

Johns Hopkins turbulence data-base: www.turbulence.pha.jhu.edu

Jones, D. A. & Titi, E.S. (1993) Upper bounds on the number of determining modes, nodes, and volume elements for the Navier–Stokes equations. *Indiana Univ. Math. J.* **42**, 875–887.

Kerr, R.M. (1985) Higher order derivative correlations and the alignment of small-scale structures in isotropic numerical turbulence. *J. Fluid Mech.* **153**, 31–58.

Kerr, R.M. (2012) Dissipation and enstrophy statistics in turbulence: Are the simulations and mathematics converging? *J. Fluid Mech.* **700**, 1–4.

Kerr, R.M. (2013) Swirling, turbulent vortex rings formed from a chain reaction of reconnection events. *Phys. Fluids* **25**, 065101, 2013.

Kurien, S. & Taylor, M.A. (2005) Direct numerical simulations of turbulence: data generation and statistical analysis. *Los Alamos Science* **29**, 142–154.

Lu, L. & Doering, C.R. (2008) Limits on enstrophy growth for solutions of the three-dimensional Navier–Stokes equations. *Indiana Univ. Math. J.* **57**, 2693–2727.

Moin, P. & Mahesh, K. (1998) Direct numerical simulation: a tool for turbulence research. *Annu. Rev. Fluid Mech.* **30**, 539–578.

Olson, E., & Titi, E.S. (2003) Determining modes for continuous data assimilation in 2D turbulence. *J. Statist. Phys.* **113**, no. 5-6, 799–840.

Olson, E. & Titi, E.S. (2008) Determining Modes and Grashof Number in 2D Turbulence – A Numerical Case Study. *Theoretical and Computational Fluid Dynamics* **22**, No. 5, 327–399.

Orszag, S.A. & Patterson, G.S. (1972) Numerical simulation of three-dimensional homogeneous isotropic turbulence. *Phys. Rev. Lett.* **28**, 76–79.

Pandit, R., Perlekar P., & Ray, S.S. (2009) Statistical properties of turbulence: an overview. *Pramana - J. Phys.* **73**, 157–191.

Rogallo, R. S. 1981 Numerical experiments in homogeneous turbulence. *NASA Tech. Memo* 81315.

Robinson, J.C. (1996) The asymptotic completeness of inertial manifolds. *Nonlinearity* **9**, 1325–1340.

Schumacher, J., Eckhardt, B., & Doering, C.R. (2010) Extreme vorticity growth in Navier–Stokes turbulence. *Phys. Letts.* **A374**, 861–865.

Sreenivasan, K.R. & Antonia, R.A. (1997) The phenomenology of small-scale turbulence. *Annu. Rev. Fluid Mech.* **29**, 435–472.

Sulem P.-L. & Frisch, U. (1975) Bounds on energy flux for finite energy turbulence. *J. Fluid Mech.* **72**, 417–424.

Temam, R. (1988) *Infinite Dimensional Dynamical Systems in Mechanics and Physics. Applied Mathematical Sciences* **68**, Springer-Verlag, Berlin.

Titi, E.S. (1990) On approximate Inertial Manifolds to the Navier–Stokes equations. *J. Math. Anal. Applns* **149**, 540–557.

Yeung, P.K., Donzis, D. & Sreenivasan, K.R. (2012) Dissipation, enstrophy and pressure statistics in turbulence simulations at high Reynolds numbers. *J. Fluid Mech.* **700**, 5–15.

5

Boussinesq equations with zero viscosity or zero diffusivity: a review

Weiwei Hu, Igor Kukavica, Fei Wang, & Mohammed Ziane

Department of Mathematics,
University of Southern California,
Los Angeles, CA 90089.
`weiweihu@usc.edu, kukavica@usc.edu,`
`wang828@usc.edu, ziane@math.usc.edu`

Abstract

In the paper, we summarize some recent results on the global existence and persistence of regularity for the 2D Boussinesq equations. We consider both, the zero diffusivity and the zero viscosity cases.

5.1 Introduction

In this review paper, we consider the 2D Boussinesq equations on $\mathbb{R}^2$ given by

$$\frac{\partial u}{\partial t} - \nu \Delta u + u \cdot \nabla u + \nabla p = \begin{pmatrix} 0 \\ \rho \end{pmatrix} \tag{5.1}$$

$$\nabla \cdot u = 0 \tag{5.2}$$

$$\frac{\partial \rho}{\partial t} - \kappa \Delta \rho + u \cdot \nabla \rho = 0 \tag{5.3}$$

with the initial conditions

$$u(0) = u_0$$

$$\rho(0) = \rho_0, \tag{5.4}$$

where u is the velocity, p is the pressure, and ρ is the variation of density. The kinematic viscosity ν and the thermal diffusivity κ are nonnegative. The Boussinesq equations have been widely studied (cf. Adhikari

Published in *Recent Progress in the Theory of the Euler and Navier-Stokes Equations*, edited by James C. Robinson, José L. Rodrigo, Witold Sadowski, & Alejandro Vidal-López. ©Cambridge University Press 2016.

et al., 2014, Berselli & Spirito, 2011, Brandolese & Schonbek, 2012, Chae, 2006, Cannon & DiBenedetto, 1980, Chen & Goubet, 2009, Constantin, Lewicka, & Ryzhik, 2006, Chae & Nam, 1997, Danchin & Paicu, 2008a,b, E & Shu, 1994, Hmidi & Keraani, 2007, 2009, Hmidi, Keraani, & Rousset, 2011, Hou & Li, 2005, Kelliher, Temam, & Wang, 2011, Moffatt, 2001, Robinson, 2001, Wang, 2005). One of the outstanding open questions is the global well-posedness and regularity in the absence of the viscous effects, i.e., when $\kappa = \nu = 0$. Even the global well-posedness with partial viscosity terms, i.e., when either $\kappa > 0$ and $\nu = 0$, or $\kappa = 0$ and $\nu > 0$, has been open until recently. In fact, the global well-posedness in the case when $\kappa = 0$ and $\nu > 0$ was listed as one of the twenty-first century problems in Moffatt (2001). Here we give a review on some recent results on the well-posedness for these two cases respectively. In addition, the global well-posedness and persistence of regularity in the case $\kappa > 0$ and $\nu = 0$ are discussed in more detail.

5.2 Persistence of regularity for the Boussinesq equations with $\nu > 0$ and $\kappa = 0$

The zero diffusivity Boussinesq equations are given by

$$\frac{\partial u}{\partial t} - \nu \Delta u + u \cdot \nabla u + \nabla p = \begin{pmatrix} 0 \\ \rho \end{pmatrix} \tag{5.5}$$

$$\nabla \cdot u = 0 \tag{5.6}$$

$$\frac{\partial \rho}{\partial t} + u \cdot \nabla \rho = 0. \tag{5.7}$$

They consist of the 2D Navier-Stokes equations (Constantin & Foias, 1988, Foias, Manley & Temam, 1988, Temam, 1997, 2001) coupled with an equation for density. (For the physical interpretation cf. Chae, 2006, Doering & Gibbon, 1995, Hou & Li, 2005, Temam, 1997.)

Chae (2006) proved the global well-posedness of the problem for initial data $(u_0, \rho_0) \in H^m(\mathbb{R}^2) \times H^m(\mathbb{R}^2)$ with integer $m \geq 3$. The theorem is stated as follows.

Theorem 5.2.1. *Chae (2006) Let $\nu > 0$ be fixed, and assume that $\operatorname{div} v_0 = 0$. Let $m \in \{3, 4, \ldots\}$, and let $(u_0, \rho_0) \in H^m(\mathbb{R}^2)$. Then, there exists a unique solution (u, ρ) of the system (5.5)–(5.7) such that*

$$u \in C([0, \infty); H^m(\mathbb{R}^2))$$

and

$$\rho \in C([0,\infty); H^m(\mathbb{R}^2)) \cap L^2_{\text{loc}}([0,\infty); H^{m+1}(\mathbb{R}^2)).$$

Moreover, for each $s < m$, the solutions (v, ρ) of (5.1)–(5.3) converge to the corresponding solutions of (5.5)–(5.7) in $C([0,T]; H^s(\mathbb{R}^2))$, for any $T > 0$, as $\kappa \to 0$.

The main ingredient used in the proof was an *a priori* estimate on $\int_0^T \|\nabla\rho\|_{L^\infty}\, dt$. A similar result below but with $(v_0, \rho_0) \in H^m \times H^{m-1}$ with $m \geq 3$ an integer was obtained by Hou & Li (2005).

Theorem 5.2.2. *Hou & Li (2005) Assume that $u_0 \in H^m(\mathbb{R}^2)$ and $\rho_0 \in H^{m-1}(\mathbb{R}^2)$ where $m \geq 3$ is an integer with $\|u_0\|_{H^m}, \|\rho_0\|_{m-1} \leq M_0$. Then for any $\nu > 0$, there exists a unique solution (u, ρ) of the viscous Boussinesq system (5.5)–(5.7). Moreover, for any $T > 0$ we have*

$$\|u(t)\|_{H^m}, \|\rho(t)\|_{H^{m-1}} \leq C(\nu, T, M_0)$$

for $0 \leq t \leq T$.

Rather than estimating $\int_0^T \|\nabla\rho\|_{L^\infty}\, dt$, the authors derived a bound on $\int_0^T \|\nabla u\|_{L^\infty}\, dt$ which in turn provides a control for $\|u\|_{H^m} + \|\rho\|_{H^{m-1}}$. We note that the Brézis–Wainger inequality

$$\|f\|_{L^\infty} \leq C\|\nabla f\|_{L^2}\left(1 + \log(1 + \|\nabla f\|_{L^p})\right)^{1/2} + C\|f\|_{L^2},$$

valid for $f \in L^2(\mathbb{R}^2) \cap W^{1,p}(\mathbb{R}^2)$ with $p > 2$, played a crucial role in both works (we refer to the reader to the paper by Brézis & Wainger, 1980).

The well-posedness in the class $H^s \times H^{s-1}$ was obtained by Danchin & Paicu (2008b) for all real $s > 2$ and for $s = 1$ by Danchin & Paicu (2008a,b). An elementary proof in the case $s = 1$ was provided by Larios, Lunasin, & Titi (2013). This left the question of persistence open for the range $1 < s \leq 2$. (By persistence we mean that the solution is unique and remains in the same space as the initial data with the norm bounded on compact time intervals.) Meanwhile, Lai, Pan, & Zhao (2011) also considered the case with zero diffusivity in a bounded domain, where the additional difficulty is the absence of the enstrophy dissipation. They obtained a unique classical solution with H^3 initial data for both the velocity and the density.

Recently, Hu, Kukavica, & Ziane (2013) showed the global existence in the case with zero diffusivity for the borderline case of the initial velocity in $H^2(\Omega)$ and the initial density in $H^1(\Omega)$ set in a smooth open bounded domain $\Omega \subseteq \mathbb{R}^2$. The main result is as follows.

Theorem 5.2.3. *Hu et al. (2013) For $u_0 \in H^2(\Omega) \cap V$ and $\rho_0 \in H^1(\Omega)$, there exists a unique global solution (u, ρ) such that*

$$u \in L^\infty\left([0, \infty); H^2(\Omega)\right) \cap L^2_{\text{loc}}\left([0, \infty); H^3(\Omega)\right)$$

and $\rho \in L^\infty\left([0, \infty); H^1(\Omega)\right)$.

For the definition of the space V, see Constantin & Foias (1988), Temam (2001).

In contrast to the approach of Lai et al. (2011), the method does not involve the time derivatives of solutions. Next, we sketch the proof. Without loss of generality, set $\nu = 1$. Standard energy estimates show that

$$\frac{\mathrm{d}}{\mathrm{d}t}\|Au\|_{L^2}^2 + \|A^{3/2}u\|_{L^2}^2 \le C\|\nabla\rho\|_{L^2}^2 + C\|u\|_{L^2}\|Au\|_{L^2}^3 \tag{5.8}$$

and

$$\frac{\mathrm{d}}{\mathrm{d}t}\|\nabla\rho\|_{L^2}^2 \le C\|\nabla u\|_{L^\infty}\|\nabla\rho\|_{L^2}^2, \tag{5.9}$$

where $A = -P\Delta$ and P is the Leray projector. Applying the Brézis–Gallouet inequality

$$\|\nabla u\|_{L^\infty} \le C\|Au\|_{L^2}\left(1 + \log\frac{\|A^{3/2}u\|_{L^2}^2}{\lambda_1\|Au\|_{L^2}^2}\right)^{1/2}, \tag{5.10}$$

which can be found in the work by Brézis & Gallouet (1980), to the right side of (5.9), we obtain an estimate

$$\frac{1}{2}\frac{\mathrm{d}}{\mathrm{d}t}\|\nabla\rho\|_{L^2}^2 \le \|Au\|_{L^2}\left(1 + \log\frac{\|A^{3/2}u\|_{L^2}^2}{\lambda_1\|Au\|_{L^2}^2}\right)^{1/2}\|\nabla\rho\|_{L^2}^2. \tag{5.11}$$

In order to conclude the proof, we use the following version of Gronwall lemma.

Lemma 5.2.4. *Hu et al. (2013) Assume that nonnegative and differentiable functions X, Y, and Z satisfy*

$$\frac{\mathrm{d}}{\mathrm{d}t}Z \le a_2 X^{1/2} Z\left(1 + \log\frac{Y}{a_3 X}\right)^{1/2} \tag{5.12}$$

and

$$\frac{\mathrm{d}}{\mathrm{d}t}X + a_1 Y \le a_2 Z + a_2 X^{3/2} \tag{5.13}$$

for some positive constants a_1, a_2, and a_3. Further assume that X is *locally integrable on* $[0, \infty)$. *Then* $X(t)$ *and* $Z(t)$ *are bounded on every interval* $[0, T]$ *with a bound depending on* $a_1, a_2, a_3, X(0), Z(0), T$, *and* $\int_0^T X(\tau) \, d\tau$.

Applying Lemma 5.2.4 with $X = \|Au\|_{L^2}^2$, $Y = \|A^{3/2}u\|_{L^2}^2$, and $Z = \|\nabla \rho\|_{L^2}^2$ to the inequalities (5.8) and (5.11) gives the desired result.

The persistence of regularity for the range $2 < s < 3$ also follows using the same approach. The remaining range $1 < s < 2$ has been addressed by Hu, Kukavica, & Ziane (2015). The main challenge is to deal with the term

$$\int \Lambda^{s-1}(u_j \partial_j \rho) \Lambda^{s-1} \rho \tag{5.14}$$

where $\Lambda = (-\Delta)^{1/2}$, needed in the H^{s-1} estimate for the density equation. The quantity (5.14) is dealt with using the following commutator inequality, improving the one provided by Chae et al. (2012).

Lemma 5.2.5. *Let* $0 < s < 1$. *For* $j = 1, 2$, *we have*

$$\|[\Lambda^s \partial_j, u]\rho\|_{L^2(\mathbb{R}^2)} \leq C \|\rho\|_{H^s(\mathbb{R}^2)} \|u\|_{H^2(\mathbb{R}^2)} \left(1 + \log \frac{\|u\|_{H^{2+s}(\mathbb{R}^2)}}{\|\rho\|_{H^2(\mathbb{R}^2)}} \right)^{1/2} \tag{5.15}$$

where C *is a positive constant depending on* s.

Using Lemma 5.2.5 and employing the technique outlined for $s = 2$ above, we obtain the following statement.

Theorem 5.2.6. *Let* $1 < s < 2$. *Assume that* $u_0 \in H^s(\mathbb{R}^2)$ *is divergence free and that* $\rho_0 \in H^{s-1}$. *Then there exists a unique global solution* (u, ρ) *such that* $u \in L^\infty\left([0, \infty); H^s(\mathbb{R}^2)\right) \cap L^2_{loc}\left([0, \infty); H^{s+1}(\mathbb{R}^2)\right)$ *and* $\rho \in L^\infty\left([0, \infty); H^{s-1}(\mathbb{R}^2)\right)$.

Here we provide a sketch of the proof. First consider the H^s estimate for the velocity equation,

$$\frac{d}{dt}\|u\|_{H^s}^2 + \|u\|_{H^{1+s}}^2 \leq f(\|u\|_{L^2}, \|\nabla u\|_{L^2}) + C\|\Lambda^{s-1}\rho\|_{L^2}^2 + \|\rho\|_{L^2}^2, \tag{5.16}$$

where f is an explicit polynomial. Using $\int u_j \Lambda^{s-1} \partial_j \rho \Lambda^{s-1} \rho = 0$, we obtain the H^{s-1} estimate for the density equation which reads

$$\frac{1}{2}\frac{d}{dt}\|\Lambda^{s-1}\rho\|_{L^2}^2 = -\int \left(\Lambda^{s-1}(u_j\partial_j\rho) - u_j\Lambda^{s-1}\partial_j\rho\right)\Lambda^{s-1}\rho \qquad (5.17)$$

$$= -\int [\Lambda^{s-1}\partial_j, u_j]\rho\Lambda^{s-1}\rho$$

$$\leq \|[\Lambda^{s-1}\partial_j, u_j]\|_{L^2}\|\Lambda^{s-1}\rho\|_{L^2}. \qquad (5.18)$$

Using (5.15), the inequality (5.18) becomes

$$\frac{d}{dt}\|\Lambda^{s-1}\rho\|_{L^2}^2 \leq C(\|\Lambda^{s-1}\rho\|_{L^2}^2 + 1)\|u\|_{H^2}\left(1 + \log_+\frac{\|u\|_{H^{1+s}}^2}{\|u\|_{H^2}^2}\right)^{1/2} \qquad (5.19)$$

where $\log_+ a = \max\{0, \log a\}$. Note that the H^1 estimate for the velocity gives $\int_0^T \|u\|_{H^2}^2\, dt < C$ for all $T > 0$. It remains to use the following Gronwall-type lemma.

Lemma 5.2.7. *Assume that differentiable functions X, $\widetilde{X}$, Y, and A satisfy*

$$\frac{d}{dt}X + \widetilde{X} \leq MY \qquad (5.20)$$

and

$$\frac{d}{dt}Y \leq MYA^{1/2}\left(1 + \log_+\frac{\widetilde{X}}{A}\right)^{1/2} \qquad (5.21)$$

where $X(0)$, $Y(0) \leq M$ for a positive constant M. Also assume that $\int_0^T A(\tau)\, d\tau \leq M$ for every $T > 0$. Then $X(t)$ and $Y(t)$ are bounded on every interval $[0, T]$ with a bound depending on M, $X(0)$, $Y(0)$, and T.

Let $X = \|u\|_{H^s}^2$ and $Y = \|\Lambda^{s-1}\rho\|_{L^2}^2 + 1$ with $\widetilde{X} = \|u\|_{H^{1+s}}^2$ and $A = \|u\|_{H^2}^2$. Applying Lemma 5.2.5 to the inequalities (5.16) and (5.19) gives the $H^s \times H^{s-1}$ persistence of regularity with $1 < s < 2$.

In addition, the proof to show the $H^s \times H^s$ persistence of regularity with $s > 1$ is also provided by Hu et al. (2015) using the same methods.

5.3 Persistence of regularity for Boussinesq equations with $\nu = 0$ and $\kappa > 0$

The zero viscosity Boussinesq equations are given by

$$\frac{\partial u}{\partial t} + u \cdot \nabla u + \nabla p = \begin{pmatrix} 0 \\ \rho \end{pmatrix} \qquad (5.22)$$

$$\nabla \cdot u = 0 \qquad (5.23)$$

$$\frac{\partial \rho}{\partial t} - \kappa \Delta \rho + u \cdot \nabla \rho = 0. \qquad (5.24)$$

For this system, Chae (2006) proved the global well-posedness for initial data $(u_0, \rho_0) \in H^m(\mathbb{R}^2) \times H^m(\mathbb{R}^2)$ with integer $m \geq 3$.

Theorem 5.3.1. *Chae (2006) Assume that* div $u_0 = 0$. *Let* $\kappa > 0$ *be fixed,* $m \geq 3$ *an integer, and assume that* $(u_0, \rho_0) \in H^m(\mathbb{R}^2) \times H^m(\mathbb{R}^2)$. *Then, there exists a unique solution* (u, ρ) *of the system* (5.22)–(5.24) *with* $\rho \in C([0, \infty); H^m(\mathbb{R}^2))$ *and*

$$u \in C([0, \infty); H^m(\mathbb{R}^2)) \cap L^2_{\mathrm{loc}}([0, T); H^{m+1}(\mathbb{R}^2)).$$

Moreover, for each $s < m$, *the solutions* (u, ρ) *of* (5.1)–(5.3) *converge to the corresponding solutions of* (5.5)–(5.7) *in* $C([0, T]; H^s(\mathbb{R}^2))$, *for any* $T > 0$, *as* $\nu \to 0$.

This result was improved in Besov spaces by Hmidi & Keraani (2009) and reads as follows.

Theorem 5.3.2. *Hmidi & Keraani (2009) Let* $q \in [0, \infty)$ *and* $r \in (2, \infty)$. *Assume that* $u_0 \in B_{q,1}^{1+2/q}$ *is a divergence-free vector field on* $\mathbb{R}^2$, *and let* $\rho_0 \in B_{q,1}^{-1+2/q} \cap L^r$, *with* $2 < r < \infty$. *There exists a unique global solution* (u, ρ) *of the Boussinesq system* (5.22)–(5.24), *such that*

$$u \in C((0, \infty); B_{q,1}^{1+2/q}(\mathbb{R}^2))$$

and

$$\rho \in L^\infty_{\mathrm{loc}}((0, \infty); B_{q,1}^{-1+2/q} \cap L^r) \cap \widetilde{L}^1_{\mathrm{loc}}\left((0, \infty); B_{q,1}^{1+2/q} \cap B_{r,\infty}^2\right).$$

For the definition of the above spaces, see Hmidi & Keraani (2007, 2009).

Li & Xu (2013) obtained the global well-posedness for initial data $(u_0, \rho_0) \in H^s(\mathbb{R}^2) \times H^s(\mathbb{R}^2)$ for real $s \geq 3$. Again the main idea of the proof is to establish an *a priori* estimate on

$$\int_0^T \left(\|\nabla u\|_{L^\infty}^2 + \|\nabla \rho\|_{L^\infty}^2 \right) \mathrm{d}t < \infty$$

for any $T > 0$.

Here we show the following result.

Theorem 5.3.3. *Let* $0 < s < 1$. *Assume that* $u_0 \in H^{2+s}(\mathbb{R}^2)$ *is divergence free and let* $\rho_0 \in H^{2+s}(\mathbb{R}^2)$. *Then there exists a unique global solution* (u, ρ) *such that*

$$u \in L_{\mathrm{loc}}^\infty \left([0, \infty); H^{2+s}(\mathbb{R}^2) \right) \tag{5.25}$$

and

$$\rho \in L_{\mathrm{loc}}^\infty \left([0, \infty); H^{2+s}(\mathbb{R}^2) \right) \cap L_{\mathrm{loc}}^2 \left([0, \infty); H^{3+s}(\mathbb{R}^2) \right). \tag{5.26}$$

Proof of Theorem 5.3.3. Without loss of generality, set $\kappa = 1$. First, taking the inner product of the density equation (5.24) with ρ yields

$$\frac{1}{2} \frac{\mathrm{d}}{\mathrm{d}t} \|\rho\|_{L^2}^2 + \|\Lambda \rho\|_{L^2}^2 = -\int_{\mathbb{R}^2} (u \cdot \nabla \rho) \rho \, \mathrm{d}x. \tag{5.27}$$

Therefore, we have

$$\|\rho(t)\|_{L^2} \le \|\rho_0\|_{L^2} \tag{5.28}$$

and

$$\int_0^t \|\Lambda \rho\|_{L^2}^2 \, \mathrm{d}x \le \frac{1}{2} \|\rho_0\|_{L^2}^2, \qquad t \in [0, T]. \tag{5.29}$$

Next, taking the inner product of the velocity equation (5.22) with u gives

$$\frac{1}{2} \frac{\mathrm{d}}{\mathrm{d}t} \|u\|_{L^2}^2 \le \|\rho\|_{L^2} \|u\|_{L^2} \tag{5.30}$$

from where

$$\frac{\mathrm{d}}{\mathrm{d}t} \|u\|_{L^2} \le \|\rho\|_{L^2} \tag{5.31}$$

and thus

$$\|u(t)\|_{L^2} \le C(T, \|\rho_0\|_{L^2}, \|u_0\|_{L^2}), t \le T. \tag{5.32}$$

Continuing as in Chae (2006), we obtain

$$\|u\|_{W^{2,r}} \le C(T, \|u_0\|_{H^{2+s}}, \|\rho_0\|_{H^{2+s}}), \ t \in [0, T], \ 2 \le r \le \frac{2}{1-s} \tag{5.33}$$

and

$$\|\rho\|_{W^{2,r}} \leq C(T, \|u_0\|_{H^{2+s}}, \|\rho_0\|_{H^{2+s}}), \ t \in [0,T], \ 2 \leq r \leq \frac{2}{1-s}. \quad (5.34)$$

Note that the restriction for r results from the embedding $H^s \hookrightarrow L^r$. Denote the vorticity by $\omega = \text{curl}\, u = \partial_1 u_2 - \partial_2 u_1$. Taking the inner product of the vorticity equation

$$\frac{\partial \omega}{\partial t} + u \cdot \nabla \omega = \text{curl}\,(\rho e_2) \quad (5.35)$$

with $\Lambda^{2+2s}\omega = -\Delta\Lambda^{2s}\omega$ gives

$$\frac{1}{2}\frac{d}{dt}\|\nabla\Lambda^s\omega\|_{L^2}^2 = -\int \partial_k\Lambda^s(u_j\partial_j\omega)\partial_k\Lambda^s\omega + \int \Lambda^{1+s}\text{curl}\,(\rho e_2) \cdot \Lambda^{1+s}\omega$$

$$= -\int \Lambda^s(\partial_k u_j\partial_j\omega)\partial_k\Lambda^s\omega - \int \Lambda^s(u_j\partial_k\partial_j\omega)\partial_k\Lambda^s\omega$$

$$+ \int \Lambda^{1+s}\text{curl}\,(\rho e_2) \cdot \Lambda^{1+s}\omega$$

$$= I_1 + I_2 + I_3, \quad (5.36)$$

where the domains of integrals are understood to be over $\mathbb{R}^2$. In order to bound I_1, we apply the fractional product rule and Brézis–Gallouet inequality and obtain

$$I_1 \leq \|\Lambda^s(\partial_k u_j\partial_j\omega)\|_{L^2}\|\partial_k\Lambda^s\omega\|_{L^2}$$

$$\leq C\big(\|\Lambda^{1+s}u\|_{L^{2/s}}\|\Lambda\omega\|_{L^{2/(1-s)}} + \|\Lambda u\|_{L^\infty}\|\Lambda^{1+s}\omega\|_{L^2}\big)\|\Lambda^{1+s}\omega\|_{L^2}$$

$$\leq C\Bigg(\|\Lambda^{1+s}u\|_{H^{1-s}}\|\Lambda\omega\|_{H^s}$$

$$+ C\|\Lambda^2 u\|_{L^2}\left(1 + \log\frac{\|u\|_{H^{2+s}}^2}{\|u\|_{H^2}^2}\right)^{1/2}\|\Lambda^{1+s}\omega\|_{L^2}\Bigg)\|\Lambda^{1+s}\omega\|_{L^2}$$

$$\leq C\|u\|_{H^2}\left(1 + \left(1 + \log\frac{\|u\|_{H^{2+s}}^2}{\|u\|_{H^2}^2}\right)^{1/2}\right)\|u\|_{H^{2+s}}^2. \quad (5.37)$$

For the term I_2, we use $\int u_j\partial_j\partial_k\Lambda^s\omega\partial_k\Lambda^s\omega = 0$ and apply the commutator estimate (5.15) from Lemma 5.2.5 in order to get

$$I_2 = -\int \left(\Lambda^s(u_j \partial_k \partial_j \omega) - u_j \partial_j \Lambda^s \partial_k \omega \right) \partial_k \Lambda^s \omega$$

$$\leq \|[\Lambda^s \partial_j, u_j] \partial_k \omega\|_{L^2} \|\partial_k \Lambda^s \omega\|_{L^2}$$

$$\leq C \|\Lambda \omega\|_{H^s} \|u\|_{H^2} \left(1 + \log \frac{\|u\|_{H^{2+s}}^2}{\|u\|_{H^2}^2} \right)^{1/2} \|\Lambda^{1+s} \omega\|_{L^2}$$

$$= C \|u\|_{H^2} \left(1 + \log \frac{\|u\|_{H^{2+s}}^2}{\|u\|_{H^2}^2} \right)^{1/2} \|u\|_{H^{2+s}}^2. \tag{5.38}$$

For the last term I_3, we have

$$I_3 \leq \|\Lambda^{2+s} \rho\|_{L^2} \|\Lambda^{1+s} \omega\|_{L^2} \leq \frac{1}{2} \|\Lambda^{2+s} \rho\|_{L^2}^2 + C \|\Lambda^{1+s} \omega\|_{L^2}^2. \tag{5.39}$$

Therefore, replacing I_1, I_2, and I_3 in (5.36) by (5.37), (5.38), and (5.39), respectively, we get

$$\frac{d}{dt} \|\Lambda^{1+s} \omega\|_{L^2}^2 \leq C \|u\|_{H^2} \left(1 + \log \frac{\|u\|_{H^{2+s}}^2}{\|u\|_{H^2}^2} \right)^{1/2} \|u\|_{H^{2+s}}^2 + \frac{1}{2} \|\Lambda^{2+s} \rho\|_{L^2}^2. \tag{5.40}$$

Now taking an inner product of the density equation (5.24) with $\Lambda^{2+s} \rho$, we have

$$\frac{1}{2} \frac{d}{dt} \|\Lambda^{2+s} \rho\|_{L^2}^2 + \|\Lambda^{3+s} \rho\|_{L^2}^2 = -\int \Lambda^{1+s}(u_j \partial_j \rho) \cdot \Lambda^{3+s} \rho$$

$$\leq \|\Lambda^{1+s}(u_j \partial_j \rho)\|_{L^2} \|\Lambda^{3+s} \rho\|_{L^2}$$

$$\leq C \big(\|\Lambda^{1+s} u\|_{L^{2/s}} \|\Lambda \rho\|_{L^{2/(1-s)}} + \|u\|_{L^\infty} \|\Lambda^{2+s} \rho\|_{L^2} \big) \|\Lambda^{3+s} \rho\|_{L^2}$$

$$\leq C \big(\|u\|_{H^2} \|\Lambda^{1+s} \rho\|_{L^2} + \|u\|_{L^2}^{1/2} \|u\|_{H^2}^{1/2} \|\Lambda^{2+s} \rho\|_{L^2} \big) \|\Lambda^{3+s} \rho\|_{L^2}$$

$$\leq \frac{1}{2} \|\Lambda^{3+s} \rho\|_{L^2}^2 + C \|u\|_{H^2}^2 \|\Lambda^{1+s} \rho\|_{L^2}^2 + C \|u\|_{L^2} \|u\|_{H^2} \|\Lambda^{2+s} \rho\|_{L^2}^2.$$

Using the bounds (5.33) and (5.34), we obtain

$$\|\Lambda^{2+s} \rho\|_{L^2} \leq C(T, \|\rho_0\|_{H^{2+s}}, \|u_0\|_{H^{2+s}}), \qquad t \in [0, T] \tag{5.41}$$

and

$$\int_0^T \|\Lambda^{3+s} \rho\|_{L^2}^2 \, dt \leq C(T, \|\rho_0\|_{H^{2+s}}, \|u_0\|_{H^{2+s}}). \tag{5.42}$$

By (5.41), we may rewrite the inequality (5.40) as

$$\frac{\mathrm{d}}{\mathrm{d}t}\|\Lambda^{1+s}\omega\|_{L^2}^2 \le C\left(\log\left(1+\|u\|_{H^{2+s}}^2\right)\right)^{1/2}\|\omega\|_{H^{2+s}}^2 + C. \qquad (5.43)$$

Using (5.30) and a Gronwall lemma, we obtain

$$\|u(t)\|_{H^{2+s}} \le C(T,\|\rho_0\|_{H^{2+s}},\|u_0\|_{H^{2+s}}), \qquad t\in[0,T]. \qquad (5.44)$$

It remains for us to show the uniqueness of the solution. Assume that there exist $(u^{(1)},p^{(1)},\rho^{(1)})$ and $(u^{(2)},p^{(2)},\rho^{(2)})$ satisfying the equations (5.22)–(5.24). Denote $\overline{u}=u^{(1)}-u^{(2)}$, $\overline{p}=p^{(1)}-p^{(2)}$, and $\overline{\rho}=\rho^{(1)}-\rho^{(2)}$. Then we have

$$\frac{\partial\overline{u}}{\partial t}+\overline{u}\cdot\nabla u^{(1)}+u^{(2)}\cdot\nabla\overline{u}+\nabla\overline{p}=\begin{pmatrix}0\\\overline{\rho}\end{pmatrix} \qquad (5.45)$$

$$\frac{\partial\overline{\rho}}{\partial t}-\Delta\overline{\rho}+\overline{u}\cdot\nabla\rho^{(1)}+u^{(2)}\cdot\nabla\overline{\rho}=0, \quad x\in\Omega, \qquad (5.46)$$

where $\nabla\cdot\overline{u}=0$ and the initial conditions satisfy $\overline{u}(0)=0$ and $\overline{\rho}(0)=0$. First, taking the inner product of the velocity equation (5.46) with $\overline{u}$, we obtain

$$\frac{1}{2}\frac{\mathrm{d}}{\mathrm{d}t}\|\overline{u}\|_{L^2}^2 \le \|\nabla u^{(1)}\|_{L^\infty}\|\overline{u}\|_{L^2}^2+\|\overline{\rho}\|_{L^2}\|\overline{u}\|_{L^2}$$

$$\le (CM_1\|\overline{u}\|_{L^2}+\|\overline{\rho}\|_{L^2})\|\overline{u}\|_{L^2} \qquad (5.47)$$

where $M_1 = \sup_{t\in[0,T]}\|u^{(1)}(t)\|_{H^{1+s}}$ for $T>0$. On the other hand, (5.46) gives

$$\frac{1}{2}\|\overline{\rho}\|_{L^2}^2 \le \|\nabla u^{(2)}\|_{L^\infty}\|\overline{\rho}\|_{L^2}^2 \le CM_1\|\overline{\rho}\|_{L^2}^2. \qquad (5.48)$$

By adding (5.45) and (5.46) and then using the Gronwall inequality, we get $\overline{u}=0$ and $\overline{\rho}=0$, which completes the proof of uniqueness. $\qquad\square$

5.4 Persistence of regularity for Boussinesq equations in general Sobolev spaces $W^{s,q}$

Some of the results in Section 5.2 extend to the general L^q setting; however, there are certain unexpected difficulties preventing us to prove persistence except for compactly supported (or at least sufficiently decaying) data. Below we describe results by Kukavica, Wang, & Ziane (preprint) showing certain results on persistence of the regularity for

Boussinesq system (5.5)–(5.7) in the space $W^{1+s,q}(\mathbb{R}^2) \times W^{s,q}(\mathbb{R}^2)$ for $s \in (0,1)$ and $q > 2$. We prove that, for initial data

$$(u_0, \rho_0) \in W^{1+s,q}(\mathbb{R}^2) \times W^{s,q}(\mathbb{R}^2),$$

the solution satisfies

$$(u(t), \rho(t)) \in W^{1+s,q}(\mathbb{R}^2) \times W^{s,q}(\mathbb{R}^2)$$

for $t \in [0, T^*]$, where T^* depends on the initial data logarithmically. Furthermore, we obtain the global persistence of regularity for initial data $(u_0, \rho_0) \in W^{1+s,q}(\mathbb{R}^2) \times W^{s,q}(\mathbb{R}^2)$ with compact support. The main theorem is stated below.

Theorem 5.4.1. *Kukavica et al. (preprint) Let $s \in (0,1)$ and $q \in (2,\infty)$. Assume that $\|u_0\|_{W^{1+s,q}}, \|\rho_0\|_{W^{s,q}} \leq M$, for some $M > 0$, and that $\operatorname{div} u_0 = 0$. Then, there exists a unique solution (u, ρ) to the equations (5.5)–(5.7) such that*

$$(u, \rho) \in C\big([0, T^*); W^{1+s,q}(\mathbb{R}^2) \times W^{s,q}(\mathbb{R}^2)\big)$$

for some T^* depending on s, q, and the initial data.

The key step for the proof of the theorem is the following commutator estimate lemma, extending the inequality due to Kato & Ponce (1988).

Lemma 5.4.2. *Kukavica et al. (preprint) Let $s \in (0,1)$ and $f, g \in \mathcal{S}(\mathbb{R}^2)$. For $1 < q < \infty$ and $j \in \{1,2\}$, we have*

$$\|[\Lambda^s \partial_j, g] f\|_{L^q(\mathbb{R}^2)} \leq C\|f\|_{L^{r_1}} \|J^{1+s} g\|_{L^{\widetilde{r_1}}} + C\|J^s f\|_{L^{r_2}} \|Jg\|_{L^{\widetilde{r_2}}}, \quad (5.49)$$

where $r_1, \widetilde{r_1}, \widetilde{r_2} \in [q, \infty]$ and $r_2 \in [q, \infty)$ satisfy

$$1/q = 1/r_1 + 1/\widetilde{r_1} = 1/r_2 + 1/\widetilde{r_2},$$

and where C is a constant depending r_1, $\widetilde{r_1}$, r_2, $\widetilde{r_2}$, s, and q.

We also need a replacement of the inequality by Beale, Kato, & Majda (1984) for our L^p setting.

Lemma 5.4.3. *Kukavica et al. (preprint) Assume that $u \in (\mathcal{S}(\mathbb{R}^2))^2$ is a vector field satisfying $\operatorname{div} u = 0$. Let $q \geq 2$ and $rs > 2$. With $\omega = \operatorname{curl} u$, the inequalities*

$$\|\Lambda u\|_{L^\infty} \leq C\big(1 + \log(1 + \|\Lambda^s \omega\|_{L^r})\big)^{1+1/q}(1 + \|\omega\|_{L^q} + \|\nabla(|\omega|^{q/2})\|_{L^2}^{2/q}) \tag{5.50}$$

and

$$\|\omega\|_{L^\infty} \leq C\big(1 + \log(1 + \|\Lambda^s \omega\|_{L^r})\big)^{1/q}(1 + \|\omega\|_{L^q} + \|\nabla(|\omega|^{q/2})\|_{L^2}^{2/q}) \tag{5.51}$$

hold, where $C = C(r, s, q)$.

With the aid of these two lemmas, we give a brief proof of Theorem 5.4.1. Without loss of generality, we assume $\nu = 1$. Multiplying the equation (5.5) with $|u|^{q-2}u$ and integrating it with respect to x, we obtain

$$\frac{1}{q}\frac{d}{dt}\|u\|_{L^q}^q + \int (u \cdot \nabla u) \cdot u|u|^{q-2}\,dx \tag{5.52}$$

$$-\int \Delta u \cdot u|u|^{q-2}\,dx + \int \nabla p \cdot u|u|^{q-2}\,dx$$

$$= \int \rho e_2 \cdot u|u|^{q-2}\,dx. \tag{5.53}$$

Using the divergence free condition and integrating by parts, we arrive at

$$\frac{1}{q}\frac{d}{dt}\|u\|_{L^q}^q + D_0 = -\int \nabla p \cdot u|u|^{q-2}\,dx + \int \rho e_2 \cdot u|u|^{q-2}\,dx$$

$$\leq \|\nabla p\|_{L^q}\|u\|_{L^q}^{q-1} + \|\rho\|_{L^q}\|u\|_{L^q}^{q-1}, \tag{5.54}$$

where

$$D_0 = \int \partial_j u_k \partial_j u_k |u|^{q-2}\,dx + (q-2)\int (u_k \partial_j u_k)(u_l \partial_j u_l)|u|^{q-4}\,dx. \tag{5.55}$$

Taking the divergence of the equation (5.5), we get

$$\|\nabla p\|_{L^q} \leq C(\|u\|_{L^\infty}\|\nabla u\|_{L^q} + \|\rho\|_{L^q}) \tag{5.56}$$

$$\leq C(\|u\|_{L^q} + \|\omega\|_{L^q})\|\omega\|_{L^q} + C\|\rho\|_{L^q}. \tag{5.57}$$

For ρ we have

$$\frac{1}{q}\frac{d}{dt}\|\rho\|_{L^q}^q = 0. \tag{5.58}$$

Since $\rho_0 \in W^{s,q} \hookrightarrow L^r$ for $r \in I$, where

$$I = \begin{cases} [q, 2q/(2-qs)], & \text{if } qs < 2 \\ [q, \infty), & \text{otherwise}, \end{cases} \tag{5.59}$$

we have

$$\|\rho(t)\|_{L^r} = C(M_1, r), \qquad r \in I, \qquad t > 0. \tag{5.60}$$

Combining (5.54) with (5.57) and (5.60) leads to

$$\frac{1}{q}\frac{d}{dt}\|u\|_{L^q}^q + D_0 \leq C(\|u\|_{W^{1,q}}\|\omega\|_{L^q} + \|\rho\|_{L^q})\|u\|_{L^q}^{q-1} \qquad (5.61)$$

$$\leq C(\|u\|_{W^{1+s,q}}\|\omega\|_{L^q} + 1)\|u\|_{L^q}^{q-1}. \qquad (5.62)$$

Taking the curl of the equation (5.5), by the L^q energy estimate, we have

$$\frac{1}{q}\frac{d}{dt}\|\omega\|_{L^q}^q + (q-1)\left(\frac{2}{q}\right)^2\|\nabla(|\omega|^{q/2})\|_{L^2}^2 = -(q-1)\int \rho|\omega|^{q-2}\partial_1\omega\,dx. \qquad (5.63)$$

Using

$$\||\omega|^{(q-2)/2}\partial_1\omega\|_{L^2} = \frac{2}{q}\|\partial_1|\omega|^{q/2}\|_{L^2}, \qquad (5.64)$$

we obtain

$$\frac{1}{q}\frac{d}{dt}\|\omega\|_{L^q}^q \leq C\|\rho\|_{L^q}^2\|\omega\|_{L^q}^{q-2}. \qquad (5.65)$$

Noting (5.60), the above inequality implies

$$\|\omega(t)\|_{L^q} \leq C(M_0, T), \qquad t \in [0, T], \qquad (5.66)$$

and

$$\int_0^T \|\nabla(|\omega|^{q/2})\|_{L^2}^2 dt < \infty. \qquad (5.67)$$

Combining (5.54), (5.57), (5.60), and (5.66) we obtain

$$\|u(t)\|_{L^q} \leq C(M_0, M_1, T), \qquad t \in [0, T]. \qquad (5.68)$$

Applying the operator Λ^s to the equation

$$\frac{\partial\omega}{\partial t} - \Delta\omega + u\cdot\nabla\omega = \text{curl}\,(\rho e_2), \qquad (5.69)$$

multiplying it with $|\Lambda^s\omega|^{q-2}\Lambda^s\omega$, and integrating it in x, we get

$$\frac{1}{q}\frac{\partial}{\partial t}\|\Lambda^s\omega\|_{L^q}^q + (q-1)\left(\frac{2}{q}\right)^2\|\nabla(|\Lambda^s\omega|^{q/2})\|_{L^2}^2$$

$$= \int \Lambda^s\partial_1\rho|\Lambda^s\omega|^{q-2}\Lambda^s\omega\,dx - \int \Lambda^s(u\cdot\nabla\omega)|\Lambda^s\omega|^{q-2}\Lambda^s\omega\,dx$$

$$= I_1 + I_2, \qquad (5.70)$$

Integrating I_1 by parts and using Hölder's inequality, we obtain

$$I_1 = -(q-1)\int \Lambda^s(\rho e_2)|\Lambda^s\omega|^{q-2}\Lambda^s\partial_1\omega\,\mathrm{d}x \tag{5.71}$$

$$= C\|\Lambda^s\rho\|_{L^q}\|\nabla(|\Lambda^s\omega|^{q/2})\|_{L^2}\|\Lambda^s\omega\|_{L^q}^{(q-2)/2}. \tag{5.72}$$

For I_2 we use Lemma 5.4.2 and Calderón–Zygmund inequality to get

$$I_2 = -\int \left(\Lambda^s(u\cdot\nabla\omega) - u\cdot\Lambda^s\nabla\omega\right)|\Lambda^s\omega|^{q-2}\Lambda^s\omega\,\mathrm{d}x \tag{5.73}$$

$$\leq C\|J^s\omega\|_{L^q}^q(\|\omega\|_{L^\infty} + \|Ju\|_{L^\infty}). \tag{5.74}$$

Therefore, by using Young's inequality we arrive at

$$\frac{1}{q}\frac{\mathrm{d}}{\mathrm{d}t}\|\Lambda^s\omega\|_{L^q}^q + (q-1)\frac{3}{q^2}\|\nabla(|\Lambda^s\omega|^{q/2})\|_{L^2}^2$$

$$\leq C\|\Lambda^s\rho\|_{L^q}^q + C\|\Lambda^s\omega\|_{L^q}^q + C\|J^s\omega\|_{L^q}^q(\|\omega\|_{L^\infty} + \|Ju\|_{L^\infty}). \tag{5.75}$$

In order to estimate $\Lambda^s\rho$, we write

$$\frac{1}{q}\|\Lambda^s\rho\|_{L^q}^q = -\int \Lambda^s(u\cdot\nabla\rho)|\Lambda^s\rho|^{q-2}\Lambda^s\rho\,\mathrm{d}x. \tag{5.76}$$

By Lemma 5.4.2 and divergence free condition, we may get

$$\int \Lambda^s\mathrm{div}(u\rho)|\Lambda^s\rho|^{q-2}\Lambda^s\rho\,\mathrm{d}x \tag{5.77}$$

$$= \int \left(\Lambda^s\mathrm{div}(u\rho) - u\cdot\nabla\Lambda^s\rho\right)|\Lambda^s\rho|^{q-2}\Lambda^s\rho\,\mathrm{d}x$$

$$\leq C\|J^s\rho\|_{L^q}\|Ju\|_{L^\infty}\|\Lambda^s\rho\|_{L^q}^{q-1} + C\|\rho\|_{L^{r_1}}\|J^{1+s}u\|_{L^{s_1}}\|\Lambda^s\rho\|_{L^q}^{q-1}, \tag{5.78}$$

where $r_1 = 2q/(2-s)$ and $s_1 = 2q/s$. Since

$$\|J^{1+s}u\|_{L^{s_1}} \leq C(\|u\|_{L^{s_1}} + \|\Lambda^{1+s}u\|_{L^{s_1}}) \leq C\|u\|_{L^{s_1}} + C\|\Lambda^s\omega\|_{L^{s_1}}$$

$$\leq C\|u\|_{W^{1+s,q}} + C\||\Lambda^s\omega|^{q/2}\|_{L^2}^{2/q(1-\alpha)}\|\nabla|\Lambda^s\omega|^{q/2}\|_{L^2}^{2\alpha/q}$$

$$= C\|u\|_{W^{1+s,q}} + C\|\Lambda^s\omega\|_{L^q}^{1-\alpha}\|\nabla|\Lambda^s\omega|^{q/2}\|_{L^2}^{2\alpha/q}, \tag{5.79}$$

where $\alpha = 1 - s/2$, the equality (5.76) leads to

$$\frac{1}{q}\|\Lambda^s\rho\|_{L^q}^q \leq C\|Ju\|_{L^\infty}\|\rho\|_{W^{s,q}}^q + C\|u\|_{W^{1+s,q}}^q \tag{5.80}$$

$$+C\|\Lambda^s\rho\|_{L^q}^q + \frac{q-1}{q^2}\||\nabla|\Lambda^s\omega|^{q/2}\|_{L^2}^2.$$

Denoting $F(t) = \|u\|_{L^q}^q + \|\Lambda^s\omega\|_{L^q}^q + \|\rho\|_{L^q}^q + \|\Lambda^s\rho\|_{L^q}^q$, summing (5.58), (5.62), (5.75), and (5.80) up and omitting the D_0 term, we obtain

$$\frac{1}{q}F(t) + (q-1)\frac{2}{q^2}\|\nabla(|\Lambda^s\omega|^{q/2})\|_{L^2}^2$$

$$\leq C\big(1 + F(t)\big) + CF(t)(\|Ju\|_{L^\infty} + \|\omega\|_{L^\infty}). \tag{5.81}$$

By Lemma 5.4.3 we further get

$$\frac{1}{q}F(t) + (q-1)\frac{2}{q^2}\|\nabla(|\Lambda^s\omega|^{q/2})\|_{L^2}^2$$

$$\leq C\big(1 + F(t)\big) + CF(t)X(t)\big(1 + \log^{1+1/q}(1 + \|\Lambda^s\omega\|_{L^r})\big), \tag{5.82}$$

where $X(t) = 1 + \|\nabla(|\omega|^{q/2})\|_{L^2}^{2/q}$ satisfies $\int_0^T X^q(t)\,dt \leq C$ by (5.67). Finally, we conclude the theorem by the following lemma.

Lemma 5.4.4. *Assume that $F(t)$ satisfies*

$$F'(t) + D(t) \leq C(1 + F(t)) + CF(t)X(t)\big(1 + \log(1 + Y(t))\big)^\alpha \tag{5.83}$$

for some $\alpha > 0$, where $F(t)$, $X(t)$, $Y(t)$ and $D(t)$ are nonnegative continuous functions on $[0, T]$ such that $\int_0^T X^q(t)\,dt < \infty$ for some $q \geq 2$ and $Y(t) \leq D(t)/2 + F(t) + M$ for some $M > 0$. There exists $T^* > 0$ such that $F(t)$ is finite and $D(t)$ is integrable on $[0, T^*]$.

A slightly different proof gives the global persistence in the space $X = (H^{1+s} \times H^s) \cap (W^{1+s,q} \times W^{s,q})$ with the norm

$$\|(u, \rho)\|_X = \max(\|u\|_{H^{1+s}}, \|u\|_{W^{1+s,q}}, \|\rho\|_{H^s}, \|\rho\|_{W^{s,q}}).$$

Theorem 5.4.5. *Kukavica et al. (preprint) Let $s \in (0, 1)$ and $q \in [2, \infty)$. Assume that $\|(u, \rho)\|_X \leq M$ where M is an arbitrary positive constant. There exists a unique solution (u, ρ) to the equations (5.5)–(5.7) such that $(u, \rho) \in C\big([0, \infty); X\big)$.*

Next we state an immediate consequence of the above theorem.

Corollary 5.4.6. *Let* $(u_0, \rho_0) \in W^{1+s,q}(\mathbb{R}^2) \times W^{s,q}(\mathbb{R}^2)$ *be compactly supported and* s, q *be the same as the above theorem. There exists a unique solution* (u, ρ) *to the equations* (5.5)–(5.7) *such that*

$$(u, \rho) \in C\big([0, \infty); W^{1+s,q}(\mathbb{R}^2) \times W^{s,q}(\mathbb{R}^2)\big).$$

5.5 Open problems

In this section, we state some remaining open problems on the regularity of solutions to the Boussinesq system.

(i) The global regularity for the zero diffusivity and inviscid Boussinesq system, i.e. $\nu = \kappa = 0$ (this problem has been open since the original work by Chae (2006), Hou & Li (2005); only local existence is known in sufficiently regular Sobolev spaces).

(ii) Global persistence for the Boussinesq equations with zero diffusivity in $H^{1+s}(\Omega) \times H^s(\Omega)$ (or $D(A^{1+s}) \times H^s(\Omega)$) for $s \in (0,1)$, where Ω is a smooth domain in $\mathbb{R}^2$.

(iii) Global persistence in $W^{1+s,q}(\mathbb{R}^2) \times W^{s,q}(\mathbb{R}^2)$ for $s \in (0,1)$ and $q > 2$ (in Kukavica et al. (preprint), we obtain global persistence for compatly supported data).

(iv) The global persistence in $W^{1+s,q}(\mathbb{T}^2) \times W^{s,q}(\mathbb{T}^2)$, where $\mathbb{T}^2$ is the periodic domain $[-\pi, \pi]^2$ for $s \in (0,1)$ and $q > 2$.

(v) The persistence in Besov sapces.

Acknowledgments

WH was supported in part by USC 2013 Zumberge Individual Research Grant, IK and FW were supported in part by the NSF grant DMS-1311943, and MZ was supported in part by the NSF grant DMS-1109562.

References

Adhikari, D., Cao, C., Wu, J., & Xu, X. (2014) Small global solutions to the damped two-dimensional Boussinesq equations. *J. Differential Equations* **256**, no. 11, 3594–3613.

Beale, J.T., Kato, T., & Majda, A. (1984) Remarks on the breakdown of smooth solutions for the 3-D Euler equations. *Comm. Math. Phys.* **94**, no. 1, 61–66.

Berselli, L.C. & Spirito, S. (2011) On the Boussinesq system: regularity criteria and singular limits. *Methods Appl. Anal.* **18**, no. 4, 391–416.

Brandolese, L. & Schonbek, M.E. (2012) Large time decay and growth for solutions of a viscous Boussinesq system. *Trans. Amer. Math. Soc.* **364**, no. 10, 5057–5090.

Brézis, H. & Gallouet, T. (1980) Nonlinear Schrödinger evolution equations, *Nonlinear Anal.* **4**, no. 4, 677–681.

Brézis, H. & Wainger, S. (1980) A note on limiting cases of Sobolev embeddings and convolution inequalities,. *Comm. Partial Differential Equations* **5**, no. 7, 773–789.

Chae, D. (2006) Global regularity for the 2D Boussinesq equations with partial viscosity terms. *Adv. Math.* **203**, no. 2, 497–513.

Chae, D., Constantin, P., Córdoba, D., Gancedo, F., & Wu, J. (2012) Generalized surface quasi-geostrophic equations with singular velocities. *Comm. Pure Appl. Math.* **65**, no. 8, 1037–1066.

Cannon, J.R. & DiBenedetto, E. (1980) The initial value problem for the Boussinesq equations with data in L^p, in *Approximation methods for Navier-Stokes problems* (Proc. Sympos., Univ. Paderborn, Paderborn, 1979), Lecture Notes in Math. **771**, Springer, Berlin, 129–144.

Constantin, P. & Foias, C. (1988) *Navier-Stokes equations*, Chicago Lectures in Mathematics, University of Chicago Press, Chicago, IL.

Chen, M. & Goubet, O. (2009) Long-time asymptotic behavior of two-dimensional dissipative Boussinesq systems. *Discrete Contin. Dyn. Syst. Ser. S* **2**, no. 1, 37–53.

Constantin, P. Lewicka, M., & Ryzhik, L. (2006) Travelling waves in two-dimensional reactive Boussinesq systems with no-slip boundary conditions. *Nonlinearity* **19**, no. 11, 2605–2615.

Chae, D. & Nam, H.S. (1997) Local existence and blow-up criterion for the Boussinesq equations. *Proc. Roy. Soc. Edinburgh Sect. A* **127**, no. 5, 935–946.

Danchin, R. & Paicu, M. (2008a) Les théorèmes de Leray et de Fujita-Kato pour le système de Boussinesq partiellement visqueux. *Bull. Soc. Math. France* **136**, no. 2, 261–309.

Danchin, R. & Paicu, M. (2008b) Les théorèmes de Leray et de Fujita-Kato pour le système de Boussinesq partiellement visqueux. *Bull. Soc. Math. France* **136**, no. 2, 261–309.

E, W. & Shu, C.W. (1994) Small-scale structures in Boussinesq convection. *Phys. Fluids* **6**, no. 1, 49–58.

Foias, C., Manley, O., & Temam, R. (1988) Modelling of the interaction of small and large eddies in two-dimensional turbulent flows. *RAIRO Modél. Math. Anal. Numér.* **22**, no. 1, 93–118.

Doering, C.R., & Gibbon, J.D. (1995) *Applied analysis of the Navier-Stokes equations*. Cambridge Texts in Applied Mathematics, Cambridge University Press, Cambridge.

Hmidi, T. & Keraani, S. (2007) On the global well-posedness of the two-dimensional Boussinesq system with a zero diffusivity. *Adv. Differential Equations* **12**, no. 4, 461–480.

Hmidi, T. & Keraani, S. (2009) On the global well-posedness of the Boussinesq system with zero viscosity. *Indiana Univ. Math. J.* **58**, no. 4, 1591–1618.

Hmidi, T., Keraani, S., & Rousset, F. (2011) Global well-posedness for Euler-Boussinesq system with critical dissipation. *Comm. Partial Differential Equations* **36**, no. 3, 420–445.

Hu, W., Kukavica, W., & Ziane, M. (2013) On the regularity for the Boussinesq equations in a bounded domain. *J. Math. Phys.* **54**, no. 8, 081507, 10.

Hu, W., Kukavica, W., & Ziane, M. (2015) Persistence of Regularity for the Viscous Boussinesq Equations with Zero Diffusivity. *Asymptotic Analysis* **91**, no. 2, 111–124.

Hou, T.Y., & Li, C. (2005) Global well-posedness of the viscous Boussinesq equations. *Discrete Contin. Dyn. Syst.* **12**, no. 1, 1–12.

Kato, T. & Ponce, G. (1988) Commutator estimates and the Euler and Navier-Stokes equations. *Comm. Pure Appl. Math.* **41**, no. 7, 891–907.

Kukavica, I., Wang, F., & Ziane, M. (preprint) Persistence of regularity for solutions of the Boussinesq equations in Sobolev spaces.

Kelliher, J.P., Temam, R., & Wang, X. (2011) Boundary layer associated with the Darcy-Brinkman-Boussinesq model for convection in porous media. *Phys. D* **240**, no. 7, 619–628.

Lai, M.-J., Pan, R., & Zhao, K. (2011) Initial boundary value problem for two-dimensional viscous Boussinesq equations. *Arch. Ration. Mech. Anal.* **199**, no. 3, 739–760.

Larios, A., Lunasin, E., & Titi, E.S. (2013) Global well-posedness for the 2D Boussinesq system with anisotropic viscosity and without heat diffusion. *J. Differential Equations* **255**, no. 9, 2636–2654.

Li, D. & Xu, X. (2013) Global wellposedness of an inviscid 2*D* Boussinesq system with nonlinear thermal diffusivity. *Dyn. Partial Differ. Equ.* **10**, no. 3, 255–265.

Moffatt, H.K. (2001) Some remarks on topological fluid mechanics, in *An introduction to the geometry and topology of fluid flows* (Cambridge, 2000), NATO Sci. Ser. II Math. Phys. Chem., **47**, Kluwer Acad. Publ., Dordrecht, 3–10.

Robinson, J.C. (2001) *Infinite-dimensional dynamical systems*, Cambridge Texts in Applied Mathematics, Cambridge University Press, Cambridge.

Temam, R. (1997) *Infinite-dimensional dynamical systems in mechanics and physics*. Applied Mathematical Sciences **68**, Springer-Verlag, New York.

Temam, R. (2001) *Navier-Stokes equations*, AMS Chelsea Publishing, Providence, RI. Reprint of the 1984 edition.

Wang, X. (2005) A note on long time behavior of solutions to the Boussinesq system at large Prandtl number, in *Nonlinear partial differential equations and related analysis*. Contemp. Math., **371**, Amer. Math. Soc., Providence, RI, 315–323.

6

Global regularity versus finite-time singularities: some paradigms on the effect of boundary conditions and certain perturbations

Adam Larios

Department of Mathematics,
University of Nebraska-Lincoln,
Lincoln, NE 68588–0130. USA.
`alarios@unl.edu`

Edriss S. Titi

Department of Mathematics,
Texas A&M University,
3368 TAMU, College Station, TX 77843–3368. USA.
`titi@math.tamu.edu`
&
Department of Computer Science and Applied Mathematics,
Weizmann Institute of Science,
Rehovot 76100. Israel.
`edriss.titi@weizmann.ac.il`

Abstract

In light of the question of finite-time blowup versus global well-posedness of solutions to problems involving nonlinear partial differential equations, we provide several cautionary examples to indicate that modifications to the boundary conditions or to the nonlinearity of the equations can effect whether the equations develop finite-time singularities. In particular, we aim to underscore the idea that in analytical and computational investigations of the blow-up of the three-dimensional Euler and Navier–Stokes equations, the boundary conditions may need to be taken into greater account. We also examine a perturbation of the nonlinearity by dropping the advection term in the evolution of the derivative of the solutions to the viscous Burgers equation, which leads to the development of singularities not present in the original equation, and indicates

Published in *Recent Progress in the Theory of the Euler and Navier-Stokes Equations*, edited by James C. Robinson, José L. Rodrigo, Witold Sadowski, & Alejandro Vidal-López. ©Cambridge University Press 2016.

that there is a regularizing mechanism in part of the nonlinearity. This simple analytical example corroborates recent computational observations in the singularity formation of fluid equations.

6.1 Introduction

A fundamental goal in the study of nonlinear initial boundary value problems involving partial differential equations is to determine whether solutions to a given equation develop a singularity in finite time. Resolving the issue of finite-time blow-up is important, in part because it can have bearing on the physical relevance and validity of the underlying model. However, determining the answer to this question is notoriously difficult for a wide range of equations, the 3D Navier–Stokes and Euler equations for incompressible fluid flow being perhaps the most well-known examples. Given that attacking the question directly is so challenging, many researchers have looked for other routes. One route is to try to simplify or modify the boundary conditions in an attempt to gain evidence for or against the occurrence of finite-time blow-up. A second route is to modify the equations in some way, and to study the modified equations with the hope of gaining insight into the blow-up of solutions to the original equations.

In this paper, we will examine several case studies related to such approaches. A major aim of the present work is to provide examples that demonstrate that one must be extremely cautious in generalizing claims about the blow-up of problems studied in idealized settings to claims about the blow-up of the original problem. A second aim is to demonstrate a phenomenon which has been observed computationally in the difficult setting of fluid flows in 3D, by means of a simple 1D example that is amenable to analysis; namely, that a seemingly harmless alteration (from the perspective of enstrophy balance) to the nonlinearity of a problem can cause the formation of a singularity, where no such singularity is present in the unaltered equation.

We will focus on three major cases. The first case examines the effect of replacing Dirichlet boundary conditions with periodic boundary conditions. This is often done in both analytical and numerical studies of the Navier–Stokes and Euler equations, for example. The original, physical equations come equipped with physical boundary conditions, such as Dirichlet boundary conditions in the case of the Navier–Stokes equations, for example. However, many such studies have tried to search for singularities of the solutions of the Euler equations in the setting of periodic boundary conditions (see, for example, Deng, Hou & Yu (2005), Hou (2009), Hou & Li (2008b,a), Kerr (1993); in particular, see the surveys by Gibbon (2008) and Gibbon, Bustamante, & Kerr (2008), and the references therein). With this in mind, in Section 6.3, we provide an example of an equation, namely

$$u_t - \Delta u = |\nabla u|^4, \tag{6.1}$$

that develops a singularity in finite time when Dirichlet boundary conditions are imposed, and yet is globally well-posed in the case of either periodic boundary conditions or the case where the domain is the full space (i.e. in the absence of physical boundaries). Therefore it may be the case that physical boundary conditions need to be taken into greater consideration in analytic and computational searches for blow-up of the solutions. Indeed, in a recent computational study, Hou & Luo (2013) (see also Luo & Hou, 2013, 2014a,b) observed the formation of a finite-time singularity *near the boundary* in the 3D Euler equations for axisymmetric flow confined in a physical cylinder, subject to no-normal flow boundary conditions. Notably, a new blow-up criterion for the 3D Euler equations in bounded domains, subject to no-normal flow boundary conditions, has been established in Gibbon & Titi (2013). It is worth stressing that this new criterion does not apply in the case of periodic boundary conditions nor when the domain is the whole space, i.e. in the absence of physical boundaries. For other issues regarding boundary behavior of the Navier–Stokes and Euler equations see the recent surveys by Bardos & Titi (2007, 2013) and the references therein.

The above discussion is particularly relevant due to the notion of "boundary-driven" mechanisms for possible blow-up. To illustrate how such a mechanism might work, we give a heuristic scenario in the context of the Navier–Stokes equations for fluid flow. It was shown in the celebrated work of Beale, Kato, & Majda (1984) that blow-up of the Euler equations occurs if and only if the vorticity becomes infinite (see also Beale et al. (1984), Constantin, Fefferman, & Majda (1996), Constantin & Fefferman (1993), Cao & Titi (2008, 2011), Prodi (1959), Serrin (1962) for additional blow-up criteria). Infinite vorticity would also cause the Navier–Stokes solutions to become singular. Now, in the setting of viscous incompressible fluids, physical boundaries are the source of vorticity shedding. Indeed, near the physical boundary of a fluid, the "no-slip" (Dirichlet) boundary conditions can cause the development of boundary layers, where the vorticity is large. If the viscous diffusion of the fluid velocity is sufficiently small in comparison to the advection, then large magnitudes of the gradient and the vorticity can be propagated from the boundary layer to the interior of the domain by the nonlinear advection term. The vorticity can then be further intensified by the nonlinear vorticity stretching term, which may thus lead to blow-up of the solution. Such a physical mechanism does not exist in the periodic setting, nor in the full space $\mathbb{R}^3$. It may therefore be illuminating to pay greater attention to the effect of boundary conditions in the search for the blow-up of solutions to the Navier–Stokes and Euler equations. We do not explore these ideas in greater detail as they are only meant to give motivation. Instead in Section 6.3 we examine the simpler equation (6.1), for which we can provide a definite answer.

In Section 6.4, we examine the Kuramoto–Sivashinsky equation,

$$u_t + \Delta^2 u + \Delta u + \tfrac{1}{2}|\nabla u|^2 = 0,$$

in a bounded domain with two different types of boundary conditions. The question of global well-posedness of this equation, when equipped with certain physically relevant boundary conditions, is still open. Recently, Pokhozhaev (2008) showed, by applying a different (non-physical) set of third-order boundary conditions, that a singularity develops in finite time. In contrast to this, we provide a different set of (also non-physical) third-order boundary conditions for which the equation is globally well-posed. Therefore, we maintain that it is difficult to obtain information about the blow-up or global well-posedness of an equation by altering its boundary conditions.

We note that such questions relating boundary conditions to blow-up can be highly relevant to applied and computational problems in science. Indeed, we recall here that such an issue occurred in the study of the planetary geostrophic model used in ocean dynamics. The model is derived asymptotically by keeping only the hydrostatic balance of the vertical momentum and the leading order geostrophic balance of the horizonal momentum, where the latter is damped by the friction with the continental shelf, while retaining the relevant physical boundary conditions. Cao, Titi, & Ziane (2004) observed that this model is over determined and hence is ill posed (it has more boundary conditions than needed for the underlying PDEs). This observation explains the numerical instabilities that had been observed near the boundary in simulations of this model. The resulting oscillations had proven difficult to eliminate, and were dealt with in Cao et al. (2004) by adding artificial higher-order diffusion corresponding to the additional boundary conditions in the model.

It is commonly believed that adding hyper-viscosity into a numerical scheme enhances the stability of the underlying scheme. In Section 6.5 we provide in example which questions the validity of this claim. That is, even though the hyper-viscous term enhances the dissipation of small scales, it destroys the maximum principle, which is an essential property for the global stability in certain physical systems. In particular, we consider an equation of the form

$$\frac{\partial u}{\partial t} + \kappa(-\Delta)^{\alpha}u - \nu\Delta u + u \cdot \nabla u = 0, \quad \text{with } \alpha > 1.$$

Finally, in Section 6.6 we consider a certain type of perturbation of the nonlinearity. In particular, in the context of the Navier–Stokes or Euler equations, by removing the advection term in the vorticity formulation, several works by Hou and collaborators (Hou & Lei, 2009; Deng et al., 2005; Hou, 2009) have observed computationally that the solutions of the altered equations seem to blow up in finite time, terming this phenomenon 'advection depleting singularity'. We give an analogous simple example based on a similar alteration of the 1D viscous Burgers equation, and we show analytically that a singularity develops in finite time, which adds credence to the numerical observations of the aforementioned works. Indeed, since the viscous Burgers equation is globally well-posed, the development of a singularity in the altered model indicates that the removed portion of the nonlinearity has a regularizing effect. However, it is worth stressing that these alterations turn out to be non-local in nature, and in the context of the hydrodynamics equations they translate to modification in the representation of the pressure term.

Many of the results and proofs are not completely new, but, for the sake of being somewhat self-contained, are collected, compared, and contrasted here. We also aim to state specific, as opposed to general results, whenever doing so simplifies the exposition. The reason for this approach is that our goal is to lay out a simple set of examples and counterexamples for the use of the reader in considering potential mechanisms for singularity formulation or prevention, in particular, in computational studies.

6.2 Preliminaries

In this section, we set some notation and recall basic results used below. We denote by L^p, $W^{s,p}$ the usual Lebesgue and Sobolev spaces. We denote by C, C', C_Ω, ... generic constants which may vary from line to line.

We recall some basic facts about the Laplace operator $\Delta := \sum_{i=1}^{n} \partial_{x_i}^2$ in the setting of either periodic or homogeneous Dirichlet boundary conditions (for proof and further discussion see Evans (2010), for example). Recall that the operator $(-\Delta)^{-1}$, subject to the appropriate boundary conditions, is a positive-definite, self-adjoint, compact operator from L^2 into itself, and therefore L^2 has an orthonormal basis that consists of eigenfunctions $\{\varphi_k\}_{k=1}^{\infty}$ (which are also eigenfunctions of $-\Delta$), corresponding to a sequence of positive eigenvalues. Since the eigenvalues of $(-\Delta)^{-1}$ can be ordered to be non-increasing, we can label the eigenvalues of $-\Delta$, which we denote by λ_k, to be such that $0 < \lambda_1 \leq \lambda_2 \leq \cdots$.

We will pay special attention to the first eigenfunction of $-\Delta$, subject to homogeneous Dirichlet boundary conditions, namely φ_1, corresponding to λ_1. We recall Hopf's Lemma, which states that $-\frac{\partial \varphi_1}{\partial \nu} > 0$ on $\partial\Omega$, where ν is the outward-pointing normal of Ω. It can also be shown that φ_1 is strictly positive on Ω. For proofs of these facts see Evans (2010), for example.

We denote the distance function to the boundary by

$$\text{dist}(x, \partial\Omega) := \inf\{|x - y| : y \in \partial\Omega\}.$$

We recall the Lions–Magenes Lemma (see, e.g., Lions & Magenes (1972) and also Temam (2001), Lemma 3.1.2), given here in a special case.

Lemma 6.2.1 (Lions–Magenes). *Let V, H, and V' be Hilbert spaces such that $V \subset H \equiv H' \subset V'$, where V' is the dual of V. Suppose that $\mathbf{u} \in L^2((0,T);V)$ and $\frac{d}{dt}\mathbf{u} \in L^2((0,T);V')$. Then $\mathbf{u}$ is equal a.e. to a continuous function from $[0,T]$ into H, and the following equality holds in the distribution sense on $(0,T)$.*

$$\frac{d}{dt}|\mathbf{u}|^2 = 2\left\langle \frac{d}{dt}\mathbf{u}, \mathbf{u} \right\rangle. \qquad (6.2)$$

Moreover, $|\mathbf{u}|^2$ is absolutely continuous. Here, $\langle \cdot, \cdot \rangle \equiv \langle \cdot, \cdot \rangle_{V',V}$ represents the action elements in V' on elements in V.

Let $\Omega \subset \mathbb{R}^n$ be a domain that is bounded in at least one direction. For all $u \in W_0^{1,p}(\Omega)$, $p \geq 1$, the following Poincaré inequality holds

$$\|u\|_{L^p} \leq C_\Omega \|\nabla u\|_{L^p}, \qquad (6.3)$$

with $C = \lambda_1^{-1/2}$ if $p = 2$.

We next recall the Gevrey classes of spatially analytic functions.

Definition 6.2.2. We define the Gevrey classes $G_\sigma^{s/2}(\mathbb{T}^n)$ of spatially analytic functions on the torus $\mathbb{T}^n := \mathbb{R}^n/(2\pi\mathbb{Z})^n$ to be the set of all $u \in L^2(\mathbb{T}^n)$ such that $\|u\|_{G_\sigma^{s/2}(\mathbb{T}^n)} < \infty$, where

$$\|u\|_{G_\sigma^{s/2}(\mathbb{T}^n)} := \left(\sum_{\mathbf{k} \in \mathbb{Z}^n} |u_{\mathbf{k}}|^2 (1 + |\mathbf{k}|^2)^s e^{2\sigma(1+|\mathbf{k}|^2)^{1/2}} \right)^{1/2}, \qquad (6.4)$$

where $u_{\mathbf{k}}$ are the Fourier coefficients of u, and where $\sigma > 0$.

Such functions are called Gevrey regular. Note that formally setting $\sigma = 0$, we recover the usual Sobolev spaces $H^s(\mathbb{T}^n)$. Furthermore, it can be shown that for $\sigma > 0$, σ is comparable to the minimal radius of analyticity of u.

6.3 Periodic versus Dirichlet boundary conditions

Consider the Cauchy problem for the following viscous Hamilton–Jacobi equation,

$$u_t - \Delta u = |\nabla u|^p, \qquad\qquad \text{in } \Omega \times (0, T), \qquad (6.5a)$$

$$u(0) = u_0, \qquad\qquad \text{in } \Omega, \qquad (6.5b)$$

equipped with either periodic boundary conditions or homogeneous Dirichlet boundary conditions. Many authors have studied the cases $p \in [0, \infty)$ (see, e.g., Alaa, 1996; Amour & Ben-Artzi, 1998; Ben-Artzi, Goodman, & Levy, 2000; Ben-Artzi, Souplet, & Weissler, 2002, 1999; Gilding, 2005; Gilding, Guedda, & Kersner, 2003), but in this work, we will focus on the case $p = 4$ for simplicity. In the case $p = 2$, the equation (6.5a) is an integrated version of the viscous Burgers equation, and is sometimes referred to as the Kardar–Parisi–Zhang equation, which is used to model the growth and roughening of certain surfaces, as derived in Kardar, Parisi, & Zhang (1986). Furthermore, (6.5) is an important test equation, since it is one of the simplest examples of a parabolic PDE with nonlinear dependence on the gradient.

In the case of periodic boundary conditions, (6.5) with $p \geq 2$ is well-posed, globally in time. However, in the Dirichlet case, and for $p > 2$, a singularity will develop in finite time, for certain initial data. We give a relatively simple proof of the well-posedness in the periodic case with $p = 4$. The proof for $p > 2$ is given in Gilding et al. (2003). For the proof of blow-up in the Dirichlet case, choosing $p = 4$ does not appear to make things significantly simpler than allowing $p > 2$, so we give the proof for $p > 2$. We follow closely the proof in Souplet (2002) to show that a singularity occurs in finite time in the Dirichlet case, at least for sufficiently large initial data in the sense given in (6.11), below.

It is worth noting that the following identity holds for sufficiently smooth functions $u = u(t, x)$:

$$(\partial_t - \Delta)e^u = (\partial_t u - \Delta u - |\nabla u|^2)e^u,$$

Thus, (6.5) can be solved explicitly in the case $p = 2$, by making the change of variables $v = e^u$ (known as the Cole–Hopf transformation for the Burgers equation), and noting that if u solves (6.5), then v solves the linear heat equation, with the corresponding boundary conditions.

6.3.1 Global well-posedness in the periodic case

We prove the global existence of solutions to (6.5), for $p = 4$ under the assumption of periodic boundary conditions. We begin by stating a special case of a theorem in Ferrari & Titi (1998) (which follows ideas from Foias & Temam, 1989), that gives short-time existence, uniqueness, and regularity (see also Takáč et al. (1996) for a different approach).

Theorem 6.3.1 (Ferrari & Titi, 1998). *Let $u_0 \in H^s(\mathbb{T}^n)$, with $s > n/2$, and $\|u_0\|_{H^s(\mathbb{T}^n)} \leq M_0$ for some $M_0 > 0$. Then there exists a $T > 0$ depending only upon M_0 such that equation (6.5), with p and positive even integer, has a unique solution u on the interval $[0, T)$ with the initial value u_0, which satisfies $u \in C([0, T); H^s(\mathbb{T}^n)) \cap L^2((0, T); H^2(\mathbb{T}^n))$, $\partial_t u \in L^2((0, T); L^2(\mathbb{T}^n))$. Moreover, $u(\cdot, t) \in G_t^{s/2}(\mathbb{T}^n)$ for $t \in [0, T)$.*

With this theorem in hand, we now state and prove a global existence theorem for (6.5) with $p = 4$ and $n = 1$. For global well-posedness in the general case, see Gilding (2005), Gilding et al. (2003).

Theorem 6.3.2. *Suppose $u_0 \in H^1(\mathbb{T})$, and consider (6.5) in the one-dimensional case with periodic boundary conditions, and $p = 4$. Then the unique (Gevrey regular) solution given by Theorem 6.3.1 can be extended to an arbitrarily large time interval $[0, T]$.*

Proof. First note that, since $p = 4$, the right-hand side of (6.5) is real analytic in u_x, and we have short-time existence and uniqueness (say, on a time interval $[0, T]$) of (6.5) under periodic boundary conditions by using, e.g., the Galerkin method. Furthermore, as shown in Ferrari

& Titi (1998), the solution is Gevrey regular in space. In particular, it has continuous derivatives of all orders. It remains to show that the solution exists globally in time. Suppose $[0, T^*)$ is the maximal interval of existence. If $T^* = \infty$ there is nothing to prove. Therefore, we assume by contradiction that $T^* < \infty$. From the above regularity, we infer in particular, that $u(\cdot, \frac{T^*}{2}) \in H^2(\mathbb{T})$. We use a technique of E. Hopf and G. Stampacchia (see Kinderlehrer & Stampacchia, 1980; Temam, 1997) to prove a maximum principle for u_x. Write $v := u_x$ and $v_*(\cdot) := u_x(\cdot, \frac{T^*}{2})$. For any function $f \in H^1$, we use the standard notation $f^+ := \max\{f, 0\}$. It is a standard exercise (see, e.g., Evans (2010), Section 5.10) to show that $f \in H^1$ implies $f^+ \in H^1$. Taking the derivative of (6.5a), we have $v_t - v_{xx} = 4v^3 v_x$. Let us define

$$\theta(x, t) := v(x, t) - \|v_*\|_{L^\infty}.$$

Since $\|v_*\|_{L^\infty}$ is a constant, $\theta_x = v_x$ and $\theta_t = v_t$, so that

$$\theta_t - \theta_{xx} - 4v^3 \theta_x = 0. \tag{6.6}$$

Taking the inner product in L^2 of (6.6) with θ^+, we integrate by parts several times and use the fact that $\theta^+ \theta = (\theta^+)^2$ to find

$$\frac{1}{2}\frac{d}{dt}\|\theta^+\|_{L^2}^2 + \|\theta_x^+\|_{L^2}^2 = \int_\mathbb{T} 4v^3 \theta_x \theta^+ \, dx = \int_\mathbb{T} 2v^3((\theta^+)^2)_x \, dx$$

$$= -\int_\mathbb{T} 6v^2 v_x(\theta^+)^2 \, dx = -\int_\mathbb{T} 2v^2((\theta^+)^3)_x \, dx$$

$$= \int_\mathbb{T} 4vv_x(\theta^+)^3 \, dx = \int_\mathbb{T} v((\theta^+)^4)_x \, dx$$

$$= -\int_\mathbb{T} v_x(\theta^+)^4 \, dx = -\int_\mathbb{T} \frac{1}{5}((\theta^+)^5)_x \, dx = 0.$$

Thus, integrating in time, for a.e. $t \in [\frac{T^*}{2}, T^*)$ we have

$$\|\theta^+(t)\|_{L^2}^2 \leq \|\theta^+(\tfrac{T^*}{2})\|_{L^2}^2 = 0.$$

Thus, $\theta^+(t) \equiv 0$, and so, $v(x, t) \leq \|v_*\|_{L^\infty}$, for $t \in [\frac{T^*}{2}, T^*)$. Similarly, one can show that $-v(x, t) \leq \|v_*\|_{L^\infty}$, and thus we have

$$\|u_x(t)\|_{L^\infty} \leq \|u_x(\tfrac{T^*}{2})\|_{L^\infty},$$

for $t \in [\frac{T^*}{2}, T^*)$. Next, taking the inner product of (6.5a) with u and using the Lions–Magenes Lemma 6.2.1, we have, for $t \in [\frac{T^*}{2}, T^*)$,

$$\frac{1}{2}\frac{d}{dt}\|u\|_{L^2}^2 + \|u_x\|_{L^2}^2 = \int_{\mathbb{T}} (u_x)^4 u \, dx \le \|u_x(\tfrac{T^*}{2})\|_{L^\infty}^4 \|u\|_{L^1}$$

$$\le C\|v_*\|_{L^\infty}^4 \|u\|_{L^2}.$$

Dropping the term $\|u_x\|_{L^2}^2$, one can conclude that

$$\|u(t)\|_{L^2(\mathbb{T})} \le CT^* \|u_x(\tfrac{T^*}{2})\|_{L^\infty}^4 + \|u(\tfrac{T^*}{2})\|_{L^2} < \infty,$$

for $t \in [\frac{T^*}{2}, T^*)$. Thus, from the above and Theorem 6.3.1 one can extend the solution beyond T^*, which leads into a contradiction. Consequently, $T^* = \infty$. $\qquad\square$

6.3.2 Finite-time blow-up in the Dirichlet case

In this section, we investigate the existence, uniqueness, and finite-time blow-up of solutions to (6.5), under the assumption of Dirichlet boundary conditions. The short-time existence and uniqueness of solutions to (6.5) can be proved by using Duhamel's principle and the Schauder fixed point theorem, see Friedman (1983), for example. We state the theorem without proof here.

Theorem 6.3.3 (Short-time existence). *Let $\Omega \subset \mathbb{R}^n$ be a bounded C^2 domain. Suppose that $u_0 \in C^{1+\alpha}(\Omega)$ for some $\alpha \in (0,1)$ and $u_0 \equiv 0$ on $\partial\Omega$. Then there exists a $T > 0$ such that (6.5) has a unique solution in $C^{1+\alpha}(\Omega \times [0,T]) \cap C(\overline{\Omega} \times [0,T])$, with $u \equiv 0$ on $\partial\Omega$.*

In Alaa (1996) it is shown that (6.5) under homogeneous Dirichlet boundary conditions cannot have a global solution if u_0 is very irregular (namely, if u_0 is a positive measure satisfying certain conditions). In Souplet (2002) it is shown that for smooth, but sufficiently large initial data (in a sense of (6.11), below), the solution u blows up in finite time for $p > 2$. In particular, it is shown that so-called "gradient blow-up" occurs: u remains uniformly bounded, but $\limsup_{t \to T^*} \|\nabla u(t, \cdot)\|_{L^\infty} = \infty$, where $T^* < \infty$ is the maximal existence time for u. The idea is to exploit properties of the first eigenvalue of the negative Laplacian operator, subject to homogeneous Dirichlet boundary conditions. Let $\lambda_1 > 0$ be the smallest eigenvalue of $-\Delta$, with homogeneous Dirichlet boundary conditions, and φ_1 a corresponding eigenfunction, chosen as in Section 6.2. We begin with two lemmas. The first is used to support the second, and the second is that $\int_\Omega (\varphi_1(x))^{-\alpha}\,dx < \infty$ for $\alpha \in (0,1)$. This means that we have a certain growth of φ_1 near the boundary, which will be needed in subsequent calculations.

Lemma 6.3.4. *Assume that $\Omega \subset \mathbb{R}^n$ is a bounded domain with C^2 boundary. Then there exists a constant $C > 0$ such that for all $x \in \Omega$,*

$$\varphi_1(x) \geq C \cdot \mathrm{dist}(x, \partial\Omega).$$

The proof is a fairly straightforward application of Hopf's Lemma $(-\partial\varphi_1/\partial\nu > 0$ on $\partial\Omega)$, but we include it here for completeness. Note that in the one-dimensional case, with $\Omega = (0,\pi)$, we have $\varphi_1(x) = \frac{\pi}{2\pi}\sin(x)$, and the results of Lemmas 6.3.4 and 6.3.5 are trivial.

Proof. Since Ω is C^2, it satisfies the interior sphere condition. Therefore for $x \in \Omega$ sufficiently close to $\partial\Omega$, we may write $x = x_0 - s\nu$ for some $x_0 \in \partial\Omega$, $s > 0$ and where ν is the exterior normal of Ω. Then $\mathrm{dist}(x, \partial\Omega) = \mathrm{dist}(x, x_0) = s$. Note that by Hopf's Lemma, $\lim_{s\to 0^+} s^{-1}\varphi_1(x_0 - s\nu) > 0$. Since $\partial\Omega$ is C^2, and Ω is bounded, we have sufficient regularity on φ_1 to conclude that this limit is uniform in x_0. Thus (choosing x closer to $\partial\Omega$ if necessary), there exists a constant $\widetilde{C} > 0$ such that $s^{-1}\varphi_1(x_0 - s\nu) \geq \widetilde{C}$, that is, $\varphi_1(x) \geq \widetilde{C} \cdot \mathrm{dist}(x, \partial\Omega)$ for all x sufficiently close to the boundary, say within an ϵ-neighborhood, for some $\epsilon \in (0,1)$. Since Ω is bounded, the set $E := \{x \in \Omega : \mathrm{dist}(x, \partial\Omega) \geq \epsilon\}$ is compact. By the elliptic maximum principle, $\varphi_1 \geq \widetilde{C}\epsilon$ on E. Setting $C = \widetilde{C}\epsilon/(1 + \mathrm{diam}(\Omega))$, we have $\varphi_1(x) \geq C \cdot \mathrm{dist}(x, \partial\Omega)$, as desired. $\square$

Using this lemma, we next show that φ_1 satisfies a certain growth condition near its zeros (i.e. near the boundary). This is the main property that is exploited by Souplet to show finite-time blow-up. A crucial lemma, proved in Souplet (2002) but also stated (without proof) in Alaa (1996) and Fila & Lieberman (1994), is the following.

Lemma 6.3.5. *Assume that Ω is a bounded domain in $\mathbb{R}^n$ with C^2 boundary, and let $\alpha \in (0,1)$. Then*

$$\int_\Omega \varphi_1^{-\alpha}(x)\,dx < \infty.$$

Proof. We follow closely the proof in Souplet (2002). We can use a partition of unity to reduce to a local argument. Since Ω is bounded with C^2 boundary, it can be locally represented as the graph of a C^2 function, say $f : U_0 \to (-\epsilon, \epsilon)$, for some $\epsilon > 0$, where $x = (x_n', x_n) \in \mathbb{R}^{n-1} \times \mathbb{R}$,

and

$$U := \{x \in \mathbb{R}^n : |x_n'| \leq \epsilon, |x_n| < \epsilon\},$$

$$U_0 := \{x_n' \in \mathbb{R}^{n-1} : |x_n'| \leq \epsilon\}, \epsilon > 0,$$

and $f(0) = 0$. Furthermore, we define

$$\omega := \{x \in U : x_n < f(x_n')\} \quad \text{and} \quad \Gamma := \{x \in U : x_n = f(x_n')\}.$$

By projecting the vector $(\vec{0}, f(x_n') - x_n)$ onto the inward-pointing normal of the graph of f, it follows that

$$\text{dist}(x, \partial\Omega) \geq \frac{f(x_n') - x_n}{\sqrt{1 + \|\nabla f\|_{L^\infty}^2}}.$$

Using this and Lemma 6.3.4, we estimate

$$\int_\omega (\varphi_1(x))^{-\alpha} \, dx \leq c^{-\alpha} \int_\omega (\text{dist}(x, \Gamma))^{-\alpha} \, dx$$

$$= c^{-\alpha} \int_{U_0} \int_{-\epsilon}^{f(x_n')} (\text{dist}(x, \Gamma))^{-\alpha} \, dx_n \, dx_n'$$

$$\leq c^{-\alpha} \left(1 + \|\nabla f\|_{L^\infty}^2\right)^{\alpha/2} \int_{U_0} \int_{-\epsilon}^{f(x_n')} (f(x_n') - x_n)^{-\alpha} \, dx_n \, dx_n' < \infty,$$

since $\alpha < 1$. $\qquad\square$

Next, we seek a lower bound on $\int_\Omega |\nabla u|^p \varphi_1(x) \, dx$.

Lemma 6.3.6. *Let $p > 2$, and suppose that $\Omega \subset \mathbb{R}^n$ is a C^2 bounded domain. Then there exists a constant $C_{\Omega,p} > 0$ such that for any function $v \in W_0^{1,p}(\Omega)$,*

$$C_{\Omega,p} \left| \int_\Omega v(t, x) \varphi_1(x) \, dx \right|^p \leq \int_\Omega |\nabla v|^p \varphi_1 \, dx. \qquad (6.7)$$

Furthermore, no such constant exists for $1 \leq p \leq 2$.

Proof. For the proof we follow closely Souplet (2002). Using Hölder's inequality, we have

$$\int_\Omega |\nabla v|\, \mathrm{d}x = \int_\Omega |\nabla v| \varphi_1^{1/p} \varphi_1^{-1/p}\, \mathrm{d}x$$

$$\le \left(\int_\Omega |\nabla v|^p \varphi_1\, \mathrm{d}x \right)^{1/p} \left(\int_\Omega \varphi_1^{-1/(p-1)}\, \mathrm{d}x \right)^{1-1/p}$$

$$\le C'_{\Omega,p} \left(\int_\Omega |\nabla v|^p \varphi_1\, \mathrm{d}x \right)^{1/p}, \qquad (6.8)$$

where, due to Lemma 6.3.5,

$$C'_{\Omega,p} := \left(\int_\Omega \varphi_1^{-1/(p-1)}\, \mathrm{d}x \right)^{1-1/p} < \infty,$$

since $p > 2$ implies $\frac{1}{p-1} \in (0,1)$.

Since Ω is bounded, we have by Poincaré's inequality (6.3), that there exists a constant C_Ω such that

$$\int_\Omega |v(x,t)|\, \mathrm{d}x \le C_\Omega \int_\Omega |\nabla v(x,t)|\, \mathrm{d}x. \qquad (6.9)$$

Therefore, (6.8) and (6.9) give

$$\left| \int_\Omega v(t,x)\varphi_1(x)\, \mathrm{d}x \right|^p \le \left(\|\varphi_1\|_{L^\infty} \int_\Omega |v(t,x)|\, \mathrm{d}x \right)^p \qquad (6.10)$$

$$\le \left(\|\varphi_1\|_{L^\infty} C_\Omega \int_\Omega |\nabla v(x,t)|\, \mathrm{d}x \right)^p$$

$$\le \left(\|\varphi_1\|_{L^\infty} C_\Omega C'_{\Omega,p} \right)^p \int_\Omega |\nabla v(x,t)|^p \varphi_1(x)\, \mathrm{d}x.$$

Setting $C_{\Omega,p} = \left(\|\varphi_1\|_{L^\infty} C_\Omega C'_{\Omega,p} \right)^{-p}$ yields (6.7).

We next give a counterexample to show that (6.7) cannot hold for all $v \in W^{1,p}(\Omega)$ in the case $p = 2$. Here, for simplicity, we only treat the one-dimensional case, with $\Omega = (0, 2\pi)$, since similar counterexamples can be constructed in higher dimensions based on the one-dimensional case, using the fact that the domain satisfies the interior sphere condition and comparing to the distance function, as in the proof of Lemma 6.3.5.

For $\epsilon \in (0, \pi/4)$, consider the function v_ϵ defined on $[0, \pi]$, given by

$$v_\epsilon(x) := \begin{cases} 0 & \text{for } x \in [0, \epsilon^2] \cup [\pi - \epsilon^2, \pi], \\ \log(x/\epsilon^2) & \text{for } x \in [\epsilon^2, \epsilon], \\ \log(1/\epsilon) & \text{for } x \in [\epsilon, \pi - \epsilon], \\ \log((\pi - x)/\epsilon^2) & \text{for } x \in [\pi - \epsilon, \pi - \epsilon^2]. \end{cases}$$

We calculate the derivative

$$v_\epsilon'(x) = \begin{cases} 0 & \text{for } x \in (0, \epsilon^2) \cup (\epsilon, \pi - \epsilon) \cup (\pi - \epsilon^2, \pi), \\ 1/x & \text{for } x \in (\epsilon^2, \epsilon), \\ 1/(x - \pi) & \text{for } x \in (\pi - \epsilon, \pi - \epsilon^2). \end{cases}$$

Notice that $v_\epsilon \in W_0^{1,2}(0, \pi)$. Furthermore, since $\sin(x) \geq \frac{2}{\pi}x$ on $[0, 2/\pi]$ and $0 < \epsilon < \pi/4$,

$$\int_0^\pi v_\epsilon(x)\sin(x)\,\mathrm{d}x = 2\int_0^{\pi/2} v_\epsilon(x)\sin(x)\,\mathrm{d}x \geq \frac{4}{\pi}\int_0^{\pi/2} v_\epsilon(x)x\,\mathrm{d}x$$

$$\geq \frac{4}{\pi}\int_\epsilon^{\pi/2} \log\left(\frac{1}{\epsilon}\right)x\,\mathrm{d}x = \frac{2}{\pi}\log\left(\frac{1}{\epsilon}\right)\left(\frac{\pi^2}{4} - \epsilon^2\right) > \frac{3\pi}{8}\log\left(\frac{1}{\epsilon}\right),$$

Thus

$$\left(\int_0^\pi v_\epsilon(x)\sin(x)\,\mathrm{d}x\right)^2 > \frac{9\pi^2}{64}\left(\log\left(\frac{1}{\epsilon}\right)\right)^2.$$

On the other hand, notice that

$$\int_0^\pi (v_\epsilon'(x))^2 \sin(x)\,\mathrm{d}x = 2\int_{\epsilon^2}^\epsilon \left(\frac{1}{x}\right)^2 \sin(x)\,\mathrm{d}x$$

$$\leq 2\int_{\epsilon^2}^\epsilon \left(\frac{1}{x}\right)^2 x\,\mathrm{d}x = 2\log\left(\frac{1}{\epsilon}\right).$$

Taking ratios of the above inequalities, we observe that

$$\frac{\left(\int_0^\pi v_\epsilon(x)\sin(x)\,\mathrm{d}x\right)^2}{\int_0^\pi (v_\epsilon'(x))^2 \sin(x)\,\mathrm{d}x} \geq \frac{\frac{9\pi^2}{64}\left(\log\left(\frac{1}{\epsilon}\right)\right)^2}{2\log\left(\frac{1}{\epsilon}\right)} = \frac{9\pi^2}{128}\log\left(\frac{1}{\epsilon}\right) \to \infty$$

as $\epsilon \to 0^+$, and therefore no finite number $C > 0$ can be chosen to make (6.7) true for all functions $v \in W_0^{1,2}(0, \pi)$. $\qquad \square$

Remark 6.3.7. The counterexample for $p = 2$ we believe to be new. A counterexample, based on a piecewise linear function, was given in the case $1 \leq p < 2$, for $n = 1$, by Bellout, Benachour, & Titi (2003).

With the above lemmas in hand, we are now ready to prove the main theorem on blow-up of (6.5) under homogeneous Dirichlet boundary conditions. As mentioned earlier, the proof is very similar to the one given in Souplet (2002), where it is also given in greater generality. The proof is given here for the sake of completeness.

Theorem 6.3.8. *Let $p > 2$ and suppose that $u_0 \in C^2(\Omega) \cap L^\infty(\Omega)$. There exists $K > 0$, given by equation (6.14) below, such that if*

$$\int_\Omega u_0(x)\varphi_1(x)\,\mathrm{d}x \geq K, \tag{6.11}$$

then any solution to (6.5), taken with homogeneous Dirichlet boundary conditions and initial data u_0, develops a singularity in finite time.

Proof. By Theorem 6.3.3, we know that there exists a time $T > 0$ and a unique solution $u \in C^{1+\alpha}([0,T] \times \Omega) \cap C([0,T] \times \overline{\Omega})$ to (6.5) satisfying $u \equiv 0$ on $\partial\Omega$. Let $T^* > 0$ be the maximal time of existence of the solution to (6.5). If $T^* < \infty$ then there is nothing to prove. Therefore, we assume by contradiction that $T^* = \infty$. Following Souplet (2002), let

$$z(t) = \int_\Omega u(t,x)\varphi_1(x)\,\mathrm{d}x. \tag{6.12}$$

The use of $z(t)$ will allow us to use standard non-existence results for ODEs, exploiting properties of φ_1. Integrating by parts, we calculate in $(0, T^*)$,

$$z'(t) + \lambda_1 z(t) = \int_\Omega u_t(t,x)\varphi_1(x)\,\mathrm{d}x - \int_\Omega u(t,x)\Delta\varphi_1(x)\,\mathrm{d}x \tag{6.13}$$

$$= \int_\Omega \big(u_t(t,x) - \Delta u(t,x)\big)\varphi_1(x)\,\mathrm{d}x = \int_\Omega |\nabla u|^p \varphi_1(x)\,\mathrm{d}x.$$

Applying Lemma 6.3.6 to (6.13) gives

$$z'(t) + \lambda_1 z(t) \geq C_{\Omega,p}(z(t))^p.$$

Now, if

$$z(0) \geq (2\lambda_1/C_{\Omega,p})^{1/(p-1)} =: K \tag{6.14}$$

then the above estimate implies that $z'(t) \geq 0$ for a short interval of time, and thus $z(t) \geq K$ for all $t \geq 0$. Let $y(t) := e^{\lambda_1 t} z(t)$. Notice that $y(0) = z(0)$. We then have

$$y'(t) \geq C_{\Omega,p} e^{\lambda_1 (1-p)t} (y(t))^p.$$

Integrating, we obtain

$$(y(0))^{1-p} - (y(t))^{1-p} \geq C_{\Omega,p} \lambda_1^{-1} (1 - e^{\lambda_1 (1-p)t}).$$

Thus,

$$(y(t))^{p-1} \geq \left((z(0))^{1-p} - C_{\Omega,p} \lambda_1^{-1} (1 - e^{\lambda_1 (1-p)t}) \right)^{-1}. \qquad (6.15)$$

Now, since $z(0) \geq K$, the right-hand side of (6.15) become infinite at some finite time $t = T^{**}$, where $e^{\lambda_1 (1-p)T^{**}} = 1/2$. Hence $T^* \leq T^{**}$, which contradicts the assumption that $T^* = \infty$. In particular, we have shown that a singularity of u develops in finite time. $\qquad \square$

Remark 6.3.9. Regarding the previous theorem, note that, by the maximum principle, since $u \equiv 0$ on $\partial\Omega$, the extreme values must occur at the initial time, so that

$$\sup_{(x,t)\in\Omega\times[0,T^*)} |u(x,t)| \leq \|u_0\|_{L^\infty} < \infty. \qquad (6.16)$$

Thus, since the solution ceases to exist after a finite time, it must do so in a norm other than L^∞.

Remark 6.3.10. In the previous theorem, the condition $p > 2$ is sharp, since (6.5) has global existence in the Dirichlet case when $p \leq 2$ (see Ferrari & Titi (1998), Gilding (2005), Gilding et al. (2003), for example). Furthermore, one can see that the reason why the proof of Lemma 6.3.8 fails is because inequality (6.7) does not hold in this case.

Let us conclude by remarking that in this section, we have seen that in the Dirichlet case, if one chooses smooth initial data, say $u_0 \in C^{1+\alpha}(\Omega)$ such that $\int_\Omega u_0 \varphi_1 \, dx$ is sufficiently large, then the solution to (6.5) will blow up in finite time. However, in the case of periodic boundary conditions, specifying that $u_0 \in H^1(\mathbb{T})$ (in fact one only needs $u_0 \in C(\mathbb{T}^n)$, as shown in Gilding et al., 2003; Gilding, 2005), the solution to (6.5) will exist globally in time. Similar results hold in full space, if one assumes, for example, that $u_0 \in C(\mathbb{R}^n) \cap L^\infty(\mathbb{R}^n)$ (see Gilding et al., 2003; Gilding, 2005). Thus, it may be the case that computational and analytic searches for blow-up in more complicated situations (such as the Navier–Stokes and Euler equations) might not provide evidence for blow-up, due

to the fact that full-space or periodic boundary conditions essentially ne-
glect any effects of the boundary, even if blow-up does occur in the case
of physical boundary conditions.

6.4 Inferences from adjusting the boundary conditions

In this section, we consider the claims that can be made with reference
to certain boundary conditions. Previously, we saw that changing from
periodic boundary conditions to physical (e.g. Dirichlet-like) boundary
conditions could determine whether or not a solution is globally well-
posed. Next, we will present an example in which a problem, given by
the Kuramoto–Sivashinsky equations, is, in the one-dimensional case,
globally well-posed under periodic, full-space, or Neumann-like boundary conditions, but loses its global regularity, in any dimension, under
another set of boundary conditions given by (6.20), below. That is to
say, in certain settings, one may engineer certain (possibly non-physical)
boundary conditions to force finite-time blow-up to occur.

We consider the Kuramoto–Sivashinsky equations, given by

$$u_t + \Delta^2 u + \Delta u + \tfrac{1}{2}|\nabla u|^2 = 0 \qquad \text{in } \Omega \times (0, T), \qquad (6.17a)$$

$$u(x, 0) = u_0(x) \qquad \text{in } \Omega. \qquad (6.17b)$$

This form of the Kuramoto–Sivashinsky equations is sometimes called
the integrated version of the Kuramoto–Sivashinsky equations (cf. equation (6.19a), below). Here, we consider $\Omega \subset \mathbb{R}^n$ to be a smooth domain.
We will discuss several variations on the boundary conditions below, as
these are the major focus of this section. Currently, even in the one-
dimensional case, the question of global existence of solutions to (6.17)
under the physical Dirichlet-like boundary conditions

$$u \equiv \Delta u \equiv 0 \text{ on } \partial\Omega, \qquad (6.18)$$

is still open. Moreover, for $n \geq 2$, the question of global well-posedness of
(6.17) in the periodic case, or in $\mathbb{R}^n$ is also an challenging open question.

As it turns out, dealing with the spatial average of the solution can
be the main obstacle in showing global regularity for (6.17), (6.19), and
to avoid this issue, many authors set $\mathbf{v} = \nabla u$, and consider instead the
differentiated version of (6.17), i.e. the system

$$\mathbf{v}_t + \Delta^2 \mathbf{v} + \Delta \mathbf{v} + (\mathbf{v} \cdot \nabla)\mathbf{v} = 0 \qquad \text{in } \Omega \times (0, T), \qquad (6.19a)$$

$$\mathbf{v}(x, 0) = \mathbf{v}_0(x) := \nabla u_0(x) \qquad \text{in } \Omega. \qquad (6.19b)$$

It is well-known that in the one-dimensional case, with either periodic ($\Omega = \mathbb{T} := \mathbb{R}/\mathbb{Z}$) or full-space ($\Omega = \mathbb{R}$) boundary conditions, (6.19) is globally well posed, and in the periodic case has a finite-dimensional global attractor and an inertial manifold (see, for example, Collet et al. (1993b), Constantin et al. (1989b,a), Foias et al. (1985b), Foias, Sell & Temam (1985a), Foias, Sell & Titi (1989), Goodman (1994), Il'yashenko (1992), Robinson (2001), Tadmor (1986), Temam (1997) and the references therein). It was shown in Cao & Titi (2006) that the only steady-state solutions to (6.17) in either $\mathbb{R}^n$ or $\mathbb{T}^n$, $n = 1, 2$, are constant functions. The question of the global well-posedness of (6.17) for $n \geq 2$ in the periodic case, or $\mathbb{R}^n$ is still open. There have been partial results in bounded domains in dimension $n \geq 2$, assuming special geometries. For instance, global well-posedness for (6.17) in dimension $n = 2, 3$ was shown in Bellout et al. (2003) for the case of radially symmetric initial data, in an annular domain $\Omega = \{x : 0 < r < |x| < R\}$, where r, R are fixed positive numbers, and the Neumann boundary conditions $\partial_r u = \partial_r \Delta u = 0$ on $\partial\Omega$ are imposed. However, the general case is currently an outstanding open problem.

One can also consider a generalization of (6.17a), namely

$$u_t + \Delta^2 u + \Delta u + \tfrac{1}{2}|\nabla u|^p = 0,$$

for some $p \geq 0$. This equation was considered in Bellout et al. (2003), where it was shown that when $p > 2$, under the boundary conditions (6.18), a singularity develops in finite time, provided that the initial data is sufficiently large in a certain sense, similar to (6.11). (In fact, in Bellout et al. (2003), the authors proved an even stronger result, as they did not need the destabilizing term Δu.) The result and proof are similar in character to that of Theorem 6.3.8, although care needs to be taken due to the fact that one no longer has a maximum principle.

Recently, in Pokhozhaev (2008), it has been shown that, for any dimension, a finite-time singularity will develop in solutions to (6.17) for a certain class of initial conditions, if one imposes the boundary conditions

$$u = 0, \qquad \frac{\partial}{\partial\nu}(u + \Delta u) = 0 \text{ on } \partial\Omega \times (0, T). \qquad (6.20)$$

The blow-up can be shown by the following calculation, which occurs in Pokhozhaev (2008) (see also Galaktionov, Mitidieri, & Pokhozhaev (2008)). Integrating equation (6.17a) in space and using the divergence theorem with boundary conditions (6.20), the Poincaré inequality, and the Cauchy–Schwarz inequality, we find that

$$\frac{\mathrm{d}}{\mathrm{d}t} \int_\Omega u \, \mathrm{d}x = \int_\Omega (-\Delta^2 u - \Delta u + \tfrac{1}{2}|\nabla u|^2) \, \mathrm{d}x = \frac{1}{2} \int_\Omega |\nabla u|^2 \, \mathrm{d}x$$

$$\geq \frac{\lambda_1}{2} \int_\Omega |u|^2 \, \mathrm{d}x \geq \frac{\lambda_1}{2|\Omega|} \left(\int_\Omega u \, \mathrm{d}x \right)^2.$$

Grönwall's inequality then yields

$$\int_\Omega u(t,x) \, \mathrm{d}x \geq \left(1 - \frac{\lambda_1 t}{2|\Omega|} \int_\Omega u_0(x) \, \mathrm{d}x \right)^{-1} \int_\Omega u_0(x) \, \mathrm{d}x.$$

Thus, if we choose the initial data such that

$$\int_\Omega u_0 \, \mathrm{d}x > 0,$$

the solution will blow up by time $T^* = \frac{2|\Omega|}{\lambda_1} \left(\int_\Omega u_0(x) \, \mathrm{d}x \right)^{-1} < \infty$.

Thus, we have seen that one can impose boundary conditions, namely conditions (6.20), to cause the solution of (6.17) to blow up for certain initial data. However, we observe again that problem (6.17) under boundary conditions (6.18) still remains open. Furthermore, we show below that if one imposes somewhat looser boundary conditions than (6.20), one can show that the solution does not blow up in finite time, at least in the one-dimensional case.

Theorem 6.4.1. *Consider the one-dimensional version of* (6.17) *on the domain* $\Omega = (0, L) \subset \mathbb{R}$, *with the boundary conditions*

$$u_x(0) = u_{xxx}(0) = u_x(L) = u_{xxx}(L) = 0. \qquad (6.21)$$

Given $T > 0$ *and* $u_0 \in H^1(0, L)$ *satisfying* (6.21), *there exists a solution* u *of* (6.17) *with* $u \in L^\infty([0,T]; L^2(0,L)) \cap L^2([0,T]; H^3(0,L))$. *Furthermore, this solution is unique, and is Gevrey regular in space for* $t > 0$.

Proof. We give only a formal existence proof here, but we remark that the proofs can be made rigorous by using the Galerkin procedure, for example. First, we notice that one can show the short-time existence by using the Galerkin procedure based on the eigenfunctions of $-\partial_{xx}$ with Neumann boundary conditions $u_x = 0$ for $x = 0, L$ (i.e. functions of the form $\cos(\pi k x/L)$). The proof of this is similar to standard proofs in periodic boundary conditions. We refer to Robinson (2001) or Temam (1997) for a demonstration of this method, and also a proof of uniqueness. Furthermore, as in Collet et al. (1993a), Foias & Temam (1989),

Ferrari & Titi (1998), Liu (1991), one can show that the solutions are Gevrey regular (analytic) in space for $t > 0$. It remains to show that the solution remains bounded for all time.

Formally taking the L^2 inner-product of (6.17a) with $-u_{xx}$ and integrating by parts, we find

$$\frac{1}{2}\frac{\mathrm{d}}{\mathrm{d}t}\|u_x\|_{L^2}^2 + \|u_{xxx}\|_{L^2}^2 = \|u_{xx}\|_{L^2}^2 + \int_0^L u_x^2 u_{xx}\,\mathrm{d}x.$$

Notice that $\int_0^L u_x^2 u_{xx}\,\mathrm{d}x = \int_0^L \frac{1}{3}\partial_x(u_x)^3\,\mathrm{d}x = 0$. Furthermore, integrating by parts and using the Cauchy–Schwarz inequality, we have

$$\|u_{xx}\|_{L^2}^2 = -\int_0^L u_x u_{xxx}\,\mathrm{d}x \leq \|u_x\|_{L^2}\|u_{xxx}\|_{L^2}. \qquad (6.22)$$

Young's inequality gives $\|u_x\|_{L^2}\|u_{xxx}\|_{L^2} \leq \frac{1}{2}\|u_x\|_{L^2}^2 + \frac{1}{2}\|u_{xxx}\|_{L^2}^2$. Combining these estimates, we have

$$\frac{\mathrm{d}}{\mathrm{d}t}\|u_x\|_{L^2}^2 + \|u_{xxx}\|_{L^2}^2 \leq \|u_x\|_{L^2}^2 \qquad (6.23)$$

Dropping the term $\|u_{xxx}\|_{L^2}^2$ and using Grönwall's inequality, we find

$$\|u_x(t)\|_{L^2}^2 \leq \mathrm{e}^t\|u_x(0)\|_{L^2}^2,$$

so that $u_x \in L^\infty([0,T]; L^2(0,L))$. Moreover, integrating (6.23) on $[0,t]$, $t \leq T$, we have

$$\|u_x(t)\|_{L^2}^2 + \int_0^t \|u_{xxx}(s)\|_{L^2}^2\,\mathrm{d}s \leq \|u_x(0)\|_{L^2}^2 + \int_0^t \|u_x\|_{L^2}^2\,\mathrm{d}s$$

$$\leq \|u_x(0)\|_{L^2}^2 + \int_0^t \mathrm{e}^s\|u_x(0)\|_{L^2}^2\,\mathrm{d}s$$

$$= \mathrm{e}^t\|u_x(0)\|_{L^2}^2.$$

Thus, $u_{xxx} \in L^2([0,T]; L^2(0,L))$. Using (6.22) and the above estimates, we find that $u_{xx} \in L^4([0,T]; L^2) \subset L^2([0,T], L^2)$. In order to show $u \in L^2([0,T], H^3)$, it remains to prove $u \in L^2([0,T], L^2)$. (Note that this does not follow directly from the Poincaré inequality, due to the boundary conditions (6.21).)

Integrating (6.17a) over $[0,L]$, we find that

$$\frac{\mathrm{d}}{\mathrm{d}t}\int_0^L u\,\mathrm{d}x = -\int_0^L u_{xxxx}\,\mathrm{d}x - \int_0^L u_{xx}\,\mathrm{d}x - \int_0^L u_x^2\,\mathrm{d}x = -\|u_x\|_{L^2}^2.$$
$$(6.24)$$

Defining $\overline{\varphi} := \int_0^L \varphi(x)\,\mathrm{d}x$, and integrating (6.24) on $[0,t]$, $t \leq T$, we have $\overline{u}(t) = \overline{u}(0) - \int_0^t \|u_x(s)\|_{L^2}^2\,\mathrm{d}s$, so that

$$|\overline{u}(t)| \leq |\overline{u}(0)| + \int_0^t \|u_x(s)\|_{L^2}^2\,\mathrm{d}s \leq |\overline{u}(0)| + (e^t - 1)\|u_x(0)\|_{L^2}^2.$$

Thus, $\overline{u} \in L^\infty([0,T])$. Now, by the Poincaré inequality,

$$\|u\|_{L^2(0,L)} - L^{1/2}|\overline{u}| = \|u\|_{L^2(0,L)} - \|\overline{u}\|_{L^2(0,L)}$$

$$\leq \|u - \overline{u}\|_{L^2(0,L)} \leq C\|u_x\|_{L^2(0,L)}.$$

Combining the above estimates, we find

$$\|u(t)\|_{L^2(0,L)} \leq L^{1/2}|\overline{u}(t)| + C\|u_x(t)\|_{L^2(0,L)}$$

$$\leq L^{1/2}\left(|\overline{u}(0)| + (e^t - 1)\|u_x(0)\|_{L^2}^2\right) + Ce^{t/2}\|u_x(0)\|_{L^2}.$$

Thus we have $u \in L^\infty([0,T]; L^2)$, and therefore from the above estimates, it follows that $u \in L^2([0,T]; H^3)$, where the bound depends only upon $|\overline{u}(0)|$, $\|u_x(0)\|_{L^2}$, T, and L. Thus, the solution can be extended globally in time. In particular, $u_x \in L^2((0,T); H^3) \subset L^2((0,T); L^\infty)$. $\square$

Remark 6.4.2. Observe that in the previous proof, one can see that $v := u_x$ satisfies (6.19) in the one-dimensional case, with $v(0) = v(L) = v_{xx}(0) = v_{xx}(L) = 0$. Furthermore, notice that, by extending v as an odd function on $[-L, L]$, this is equivalent to the case with periodic boundary conditions on $[-L, L]$, where the functions are restricted to be odd functions, where it is well known that one has global well-posedness and Gevrey regularity, as studied, for example, in Il'yashenko (1992), Tadmor (1986), Robinson (2001), Temam (1997) and the references therein.

6.5 Is hyper-viscosity stabilizing?

In many numerical simulations, especially for geophysical flows, a hyper-viscosity term of the form $(-\Delta)^\alpha$, with $\alpha > 1$, is used to stabilize the underlying numerical scheme. In the presence of physical boundaries, such as in ocean dynamics models, these artificial hyper-viscosity operators require additional *non-physical* boundary conditions. Even if we set aside this issue with the artificial boundary conditions, we are still faced with the question whether hyper-viscosity is always a stabilizing mechanism. To make our point we consider, for example, the 2D and 3D differentiated form of the viscous Burgers equations,

$$\frac{\partial \mathbf{u}}{\partial t} - \nu \Delta \mathbf{u} + (\mathbf{u} \cdot \nabla)\mathbf{u} = 0, \qquad (6.25)$$

subject to periodic boundary conditions. On the one hand, and as was observed in Section 6.3, system (6.25) is globally well-posed in dimension $n = 2, 3$ for initial data in H^s, $s > n/2 + 2$, thanks to the maximum principle, namely, $\|\mathbf{u}(\cdot, t)\|_{L^\infty} \leq \|\mathbf{u}_0\|_{L^\infty}$, for all $t \geq 0$ (see, e.g., Ladyzhenskaya, Solonnikov, & Ural'ceva (1968), Theorem 7.1, page 596). On the other hand, if one adds a hyper-viscosity term to (6.25), and considers instead

$$\frac{\partial \mathbf{u}}{\partial t} + \kappa(-\Delta)^\alpha \mathbf{u} - \nu \Delta \mathbf{u} + (\mathbf{u} \cdot \nabla)\mathbf{u} = 0, \quad \text{with } \alpha > 1, \qquad (6.26)$$

with periodic boundary conditions, nothing is known about the global regularity of (6.26) for large initial data, even in the two-dimensional case. This is because we lose the maximum principle in the hyper-viscous case, which is the only global *a priori* bound available for (6.25). Indeed, it would be interesting if one could show that (6.25) develops a finite-time singularity, while (6.25) is globally well-posed. In particular, any global regularity result concerning (6.25) will shed light on the question of global regularity for the 2D and 3D Kuramoto–Sivashinsky equation (see, e.g., Bellout et al. (2003), Cao & Titi (2006) for further discussion of this problem). We observe that in the 1D case, global regularity can be established by standard energy methods (see, e.g., Collet et al. (1993b), Constantin et al. (1989a), Goodman (1994), Tadmor (1986), Temam (1997) and references therein).

6.6 Singularity formation by altering the nonlinearity

In this section we remove the advection term in the evolution of the derivative of the Burgers equation and show that the resulting equation blows up in finite time. The alteration in nonlinearity, and the resulting blow-up phenomenon, are analogous to a phenomenon observed computationally in Deng et al. (2005), Hou (2009), Hou & Lei (2009) for much more complicated equations governing fluids. Namely, it was noticed in simulations that, by removing the advection term in the vorticity evolution of the 3D axisymmetric Navier–Stokes or Euler equations, one can seemingly cause these equations to blow up in finite time via a certain mechanism. However, when the nonlinearity is restored to its original form, the mechanism seems to disappear, and therefore, it is claimed that it is reasonable to expect that the advection is depleting the singularity.

Remarkably, a similar result was established by Constantin (1986) for the full three-dimensional Euler equations in the whole space. Specifically, he shows that for the vorticity formulation of the 3D Euler equations in the whole space without the advection term there are special solutions that blow-up in finite time. We believe that similar argument can be also applied to the vorticity formulation of the Navier–Stokes without advection.

Our purpose in this section is two-fold. First, to show analytically that this phenomenon does indeed occur, at least in the simpler setting of the 1D viscous Burgers equation. The second is to shed more light on this mechanism and to stress that it is not precisely the advection term that depletes singularity; it is rather a non-local alteration of the nonlinear and the pressure terms that cause this effect, and that, sometimes, such a non-local alteration might cause the opposite effect. In other words, we observe that this alteration of the advection term, in the evolution of the derivative in our example, and in the evolution of the vorticity in Deng et al. (2005), Hou (2009), Hou & Lei (2009), are in fact non-local in nature and are not as simple as they might seem naively. This is because any non-local change in the nonlinearity of the hydrodynamic equations is in effect an alteration in the representation of the pressure term, which is the major obstacle in the study of three-dimensional hydrodynamic equations.

To illustrate the observation made in Deng et al. (2005), Hou (2009), Hou & Lei (2009), we consider the unforced 3D Navier–Stokes equations for incompressible flow, namely

$$-\partial_t \mathbf{u} + (\mathbf{u} \cdot \nabla)\mathbf{u} = -\nabla p + \nu \Delta \mathbf{u}, \qquad (6.27a)$$

$$\nabla \cdot \mathbf{u} = 0, \qquad (6.27b)$$

in the whole space $\mathbb{R}^3$, and with a given initial condition. Here $\mathbf{u} = \mathbf{u}(x_1, x_2, x_3, t)$ is the vector-valued velocity of a fluid, and the function $p = p(x_1, x_2, x_3, t)$ is the pressure. Let us define $\boldsymbol{\omega} := \nabla \times \mathbf{u}$, which is known as the vorticity of the fluid. Taking the curl of (6.27a) and using (6.27b), we obtain the well-known vorticity equation namely,

$$\partial_t \boldsymbol{\omega} + (\mathbf{u} \cdot \nabla)\boldsymbol{\omega} = (\boldsymbol{\omega} \cdot \nabla)\mathbf{u} + \nu \Delta \boldsymbol{\omega}. \qquad (6.28)$$

Recently, in Hou & Lei (2009), a reformulation of (6.28) was given in the axisymmetric case. Furthermore, it was suggested in Hou & Lei (2009) and Hou (2009), based on numerical simulations, that in this new formulation, the analogue of the advection term $(\mathbf{u} \cdot \nabla)\boldsymbol{\omega}$ may prevent the blow-up of solutions. Specifically, it is suggested that this term may be responsible for depleting the singularity. It is worth noting that a similar phenomenon is also conjectured to occur in a generalization of the Constantin-Lax-Majda equation; see, e.g., Okamoto, Sakajo, & Wunsch (2008) and the references therein.

To parallel the above remarks about the Navier–Stokes equations, we consider the one-dimensional viscous Burgers equation with Neumann boundary conditions on the interval $(0, \pi)$:

$$u_t + uu_x = \nu u_{xx}, \quad u_x(0,t) = u_x(\pi, t) = 0, \quad u(x, 0) = u_0(x), \quad (6.29)$$

with viscosity $\nu > 0$. Cao & Titi (2002) proved that for $u_0 \in C(\overline{\Omega})$ there exists a unique (global) solution u to (6.29) satisfying

$$u \in L^2_{\text{loc}}((0, T]; H^3) \cap C((0, T]; H^2) \cap L^2((0, T]; H^1) \cap C([0, T]; L^2).$$

If $u_0 \in H^1(\Omega) \cap C(\overline{\Omega})$, one can additionally show that $u \in C([0, T]; H^1)$. (See, e.g., Ladyzhenskaya et al. (1968) for classical results on quasilinear parabolic equations, such as the viscous Burgers equation.)

Differentiating the equation in (6.29) with respect to x and setting $\omega := u_x$, we obtain

$$\omega_t + u\omega_x = \nu\omega_{xx} - \omega^2. \qquad (6.30)$$

We show that removing the advection term $u\omega_x$ from (6.30) allows for solutions that develop a singularity in finite time. We note that this phenomenon was also pointed out, without proof, in Ohkitani (2008), which is a review of Okamoto et al. (2008). Consider the problem

$$\omega_t = \nu\omega_{xx} - \omega^2, \qquad \omega\big|_{\partial\Omega} = 0, \qquad \omega(x, 0) = \omega_0(x). \qquad (6.31)$$

A generalized version of the equation in (6.31) was studied (with a positive sign on ω^2 and positive initial data) in Fujita (1966) (see also Ball, 1977, Quittner & Souplet, 2007). Since (6.31) is the viscous (PDE) version of the Riccati equation $\dot{y} = y^2$, it is not surprising that it can develop a singularity in finite time. Indeed, in Quittner & Souplet (2007) it was proven that for initial data which is everywhere positive, solutions must develop a singularity in finite time. Here, for the sake of completeness, we give a different proof, which shows that blow-up may also occur for initial data that is not everywhere positive.

Theorem 6.6.1. *There exists an $M > 0$ such that if the initial data $\omega_0 \in C([0, \pi])$ satisfies*

$$\int_0^\pi \omega_0(x) \sin(x) \, \mathrm{d}x < -M,$$

then the corresponding solution ω to (6.31) blows up in finite time. More precisely, there exists a time $T^* \in (0, \infty)$ such that

$$\lim_{t \to T^{*-}} \int_0^\pi \omega(x, t) \sin(x) \, \mathrm{d}x = -\infty. \tag{6.32}$$

Proof. We proceed somewhat formally, as we only wish to illustrate the main ideas. For notational simplicity we set $\varphi(x) = \sin(x)$. Taking the inner product of (6.31) with φ and integrating by parts twice gives

$$\frac{\mathrm{d}}{\mathrm{d}t} \int_0^\pi \omega\varphi \, \mathrm{d}x = \nu \int_0^\pi \omega\varphi_{xx} \, \mathrm{d}x - \int_0^\pi \omega^2\varphi \, \mathrm{d}x. \tag{6.33}$$

By the Cauchy–Schwarz inequality,

$$\left(\int_0^\pi \omega\varphi \, \mathrm{d}x \right)^2 = \left(\int_0^\pi \omega\varphi^{\frac{1}{2}}\varphi^{\frac{1}{2}} \, \mathrm{d}x \right)^2$$

$$\leq \left(\int_0^\pi \omega^2\varphi \, \mathrm{d}x \right) \left(\int_0^\pi \varphi \, \mathrm{d}x \right) = 2 \int_0^\pi \omega^2\varphi \, \mathrm{d}x.$$

Furthermore,

$$\int_0^\pi \omega\varphi_{xx} \, \mathrm{d}x = \int_0^\pi \omega\varphi^{\frac{1}{2}}\varphi_{xx}\varphi^{-\frac{1}{2}} \, \mathrm{d}x$$

$$\leq \left(\int_0^\pi \omega^2\varphi \, \mathrm{d}x \right)^{\frac{1}{2}} \left(\int_0^\pi \varphi_{xx}^2\varphi^{-1} \, \mathrm{d}x \right)^{\frac{1}{2}}$$

$$= \sqrt{2} \left(\int_0^\pi \omega^2\varphi \, \mathrm{d}x \right)^{\frac{1}{2}}.$$

Using these estimates in (6.33) along with Young's inequality yields

$$\frac{\mathrm{d}}{\mathrm{d}t} \int_0^\pi \omega\varphi \, \mathrm{d}x \leq \nu\sqrt{2} \left(\int_0^\pi \omega^2\varphi \, \mathrm{d}x \right)^{\frac{1}{2}} - \int_0^\pi \omega^2\varphi \, \mathrm{d}x$$

$$\leq \nu^2 - \frac{1}{2} \int_0^\pi \omega^2\varphi \, \mathrm{d}x \leq \nu^2 - \frac{1}{4} \left(\int_0^\pi \omega\varphi \, \mathrm{d}x \right)^2.$$

Setting $y(t) := \int_0^\pi \omega\varphi \, \mathrm{d}x$, we have $\dot{y} \leq \nu^2 - y^2/4$. Choosing $y(0) <$

$-\sqrt{8}\nu$, we have by continuity that $y(t) < -\sqrt{8}\nu$ for all $t \in [0, \delta]$, for some $\delta > 0$. Thus, $\dot{y} < -\frac{1}{8}y^2$ on $[0, \delta]$, so y is decreasing $[0, \delta]$, and we therefore have that $y(t) < -\sqrt{8}\nu$ for all time. Therefore, $\dot{y} < -\frac{1}{8}y^2$ for all time. Integrating on $[0, t]$, we find

$$y(t) < (\tfrac{t}{8} + (y(0))^{-1})^{-1}.$$

Thus, setting $M = -\sqrt{8}\nu$ and $T^* = -8(y(0))^{-1} = -8 \left(\int_0^\pi \omega_0 \varphi \, dx \right)^{-1} > 0$, we obtain (6.32). $\qquad\square$

We notice that in eliminating the advection term $u\omega_x$ from (6.30), we have gone from a non-local equation to a local equation, since u is the anti-derivative of ω, which is non-local in ω. Thus, perhaps the regularizing effect of the term $u\omega_x$ is due in part to its non-local nature.

An analogous effect appears in the context of the Navier–Stokes equations. Notice that dropping the advection term in (6.28) will affect the pressure term. Indeed, if we make a drastic alteration to the Navier–Stokes equations and formally drop the pressure term entirely (and also drop (6.27b), so that the system is not overdetermined), the result is the 3D viscous Burgers equation, which is known to be globally well-posed (see, e.g., Ladyzhenskaya et al., 1968). On the other hand, consider the 2D Euler equations. These equations are known to be globally well-posed, but by formally dropping the pressure term (and again the divergence-free condition), we arrive at the 2D inviscid Burgers equation, which blows up in finite time.

In conclusion any non-local alteration of the nonlinearity in the hydrodynamic equations is in effect leading to alteration in the pressure representation. This in turn might be a stabilizing or destabilizing mechanism of the modified equation.

Acknowledgements

This work is supported in part by the Minerva Stiftung Foundation, and the National Science Foundation grants numbers DMS–1009950, DMS–1109640 and DMS–1109645.

References

Alaa, N. (1996) Solutions faibles d'équations paraboliques quasilinéaires avec données initiales mesures. *Ann. Math. Blaise Pascal* **3**, no. 2, 1–15.

Amour, L. & Ben-Artzi, M. (1998) Global existence and decay for viscous Hamilton–Jacobi equations. *Nonlinear Anal.* **31**, no. 5-6, 621–628.

Ball, J.M. (1977) Remarks on blow-up and nonexistence theorems for non-linear evolution equations. *Quart. J. Math. Oxford Ser.* (2) **28**, no. 112, 473–486.

Bardos, C. & Titi, E.S. (2007) Euler equations of incompressible ideal fluids. *Uspekhi Matematicheskikh Nauk, UMN* **62**, no. 3, 5–46. Also in *Russian Mathematical Surveys* **62**, no. 3, 409–451.

Bardos, C. & Titi, E.S. (2013) Mathematics and turbulence: where do we stand? *Journal of Turbulence* **14**, no. 3, 42–76.

Beale, J.T., Kato, T. & Majda, A.J. (1984) Remarks on the breakdown of smooth solutions for the 3-D Euler equations. *Comm. Math. Phys.* **94(1)**, 61–66.

Bellout, H., Benachour, S., & Titi, E.S. (2003) Finite-time singularity versus global regularity for hyper-viscous Hamilton-Jacobi-like equations. *Nonlinearity* **16**, no. 6, 1967–1989.

Ben-Artzi, M., Souplet, P., & Weissler, F.B. (1999) Sur la non-existence et la non-unicité des solutions du problème de Cauchy pour une équation parabolique semi-linéaire. *C. R. Acad. Sci. Paris Sér. I Math.* **329**, no. 5, 371–376.

Ben-Artzi, M., Goodman, J., & Levy, A. (2000) Remarks on a nonlinear parabolic equation. *Trans. Amer. Math. Soc.* **352**, no. 2, 731–751.

Ben-Artzi, M., Souplet, P., & Weissler, F.B. (2002). The local theory for viscous Hamilton-Jacobi equations in Lebesgue spaces. *J. Math. Pures Appl.* **81**, no. 4, 343–378.

Cao, C., & Titi, E. S. (2002) Asymptotic behavior of viscous 1-D scalar con-servation laws with Neumann boundary conditions, in *Mathematics & Mathematics Education (Bethlehem, 2000)*, World Sci. Publ., River Edge, NJ, 306–324.

Cao, Y. & Titi, E.S. (2006) Trivial stationary solutions to the Kuramoto–Sivashinsky and certain nonlinear elliptic equations. *J. Differential Equations* **231**, no. 2, 755–767.

Cao, C. & Titi, E.S. (2009) Regularity criteria for the three-dimensional Navier–Stokes equations. *Indiana Univ. Math. J.* **57**, no. 6, 2643–2661.

Cao, C., & Titi, E.S. (2011) Global regularity criterion for the 3D Navier–Stokes equations involving one entry of the velocity gradient tensor. *Arch. Ration. Mech. Anal.* **202**, no. 3, 919–932.

Cao, C., Titi E.S., & Ziane, M. (2004) A "horizontal" hyper–diffusion 3-D thermocline planetary geostrophic model: well-posedness and long time behavior. *Nonlinearity* **17**, 1749–1776.

Collet, P., Eckmann, J.-P., Epstein, H., & Stubbe, J. (1993) Analyticity for the Kuramoto–Sivashinsky equation. *Phys. D* **67**, no. 4, 321–326.

Collet, P., Eckmann, J.-P., Epstein, H., & Stubbe, J. (1993) A global attract-ing set for the Kuramoto–Sivashinsky equation. *Comm. Math. Phys.* **152**, no. 1, 203–214.

Constantin, P. (1986) Note on loss of regularity for solutions of the 3-D incom-pressible Euler and related equations. *Comm. Math. Phys.* **104**, 311–326.

Constantin, P. & Fefferman, C. (1993) Direction of vorticity and the problem of global regularity for the Navier–Stokes equations. *Indiana Univ. Math. J.* **42**, no. 3, 775–789.

Constantin, P. & Foias, C. (1988) *Navier–Stokes Equations*. Chicago Lectures in Mathematics, University of Chicago Press, Chicago, IL.

Constantin, P., Fefferman, C,. & Majda, A.J. (1996) Geometric constraints on potentially singular solutions for the 3-D Euler equations. *Comm. Partial Differential Equations* **21**, no. 3–4, 559–571.

Constantin, P., Foias, C., Nicolaenko, B., & Temam, R. (1989) *Integral manifolds and inertial manifolds for dissipative partial differential equations.* Applied Mathematical Sciences **70**. Springer-Verlag, New York.

Constantin, P., Foias, C., Nicolaenko, B., & Temam, R. (1989) Spectral barriers and inertial manifolds for dissipative partial differential equations. *J. Dynam. Differential Equations* **1**, no. 1, 45–73.

Deng, J., Hou, T.Y., & Yu, X. (2005) Geometric properties and nonblowup of 3D incompressible Euler flow. *Comm. Partial Differential Equations* **30**, no. 1–3, 225–243.

Evans, L.C. (2010) *Partial differential equations.* Graduate Studies in Mathematics **19**. American Mathematical Society, Providence, RI.

Ferrari, A. & Titi, E.S. (1998) Gevrey regularity for nonlinear analytic parabolic equations. *Comm. Partial Differential Equations* **23**, no 1–2, 1–16.

Fila, M., & Lieberman, G.M. (1994) Derivative blow-up and beyond for quasilinear parabolic equations. *Differential Integral Equations* **7**, no. 3–4, 811–821.

Foias, C. & Temam, R. (1989) Gevrey class regularity for the solutions of the Navier–Stokes equations. *J. Funct. Anal.* **87**, no. 2, 359–369.

Foias, C., Sell, G.R., & Temam, R. (1985a) Variétés inertielles des équations différentielles dissipatives. *C. R. Acad. Sci. Paris Sér. I Math.* **301**, no. 5, 139–141.

Foias, C., Nicolaenko, B., Sell, G.R., & Temam, R. (1985b) Variétés inertielles pour l'équation de Kuramoto-Sivashinski. *C. R. Acad. Sci. Paris Sér. I Math.* **301**, no. 6, 285–288.

Foias, C., Sell, G.R., & Titi, E.S. (1989) Exponential tracking and approximation of inertial manifolds for dissipative nonlinear equations. *J. Dynam. Differential Equations* **1**, no. 2, 199–244.

Friedman, A. (1983) *Partial differential equations of parabolic type.* R.E. Krieger Pub. Co, Malabar.

Fujita, H. (1966) On the blowing up of solutions of the Cauchy problem for $u_t = \Delta u + u^{1+\alpha}$. *J. Fac. Sci. Univ. Tokyo Sect. I* **13**, 109–124.

Galaktionov, V.A., Mitidieri, È, & Pokhozhaev, S.I. (2008) Existence and nonexistence of global solutions of the Kuramoto–Sivashinsky equation. *Dokl. Akad. Nauk* **419**, no. 4, 439–442.

Gibbon, J.D. (2008) The three-dimensional Euler equations: where do we stand? *Phys. D* **237**, no. 14–17, 1894–1904.

Gibbon, J.D., Bustamante, M., & Kerr, R.M. (2008) The three-dimensional Euler equations: singular or non-singular? *Nonlinearity* **21**, no. 8, 123–129.

Gibbon, J.D. & Titi, E.S. (2013) 3D incompressible Euler with a passive scalar: a road to blow up? *Journal of Nonlinear Science* **23**, no. 6, 993–1000.

Gilding, B.H. (2005) The Cauchy problem for $u_t = \Delta u + |\nabla u|^q$, large-time behaviour. *J. Math. Pures Appl.* **84**, no. 6, 753–785.

Gilding, B.H., Guedda, M., & Kersner, R. (2003) The Cauchy problem for $u_t = \Delta u + |\nabla u|^q$. *J. Math. Anal. Appl.* **284**, no. 2, 733–755.

Goodman, J. (1994) Stability of the Kuramoto–Sivashinsky and related systems. *Comm. Pure Appl. Math.* **47**, no. 3, 293–306.

Hou, T.Y. (2009) Blow-up or no blow-up? A unified computational and analytic approach to 3D incompressible Euler and Navier–Stokes equations. *Acta Numer.* **18**, 277–346.

Hou, T.Y. & Lei, Z. (2009) On the stabilizing effect of convection in three-dimensional incompressible flows. *Comm. Pure Appl. Math.* **62**, no. 4, 501–564.

Hou, T.Y. & Li, R. (2008a) Blowup or no blowup? The interplay between theory and numerics. *Phys. D* **237**, no. 14–17, 1937–1944.

Hou, T.Y. & Li, R. (2008b) Numerical study of nearly singular solutions of the 3-D incompressible Euler equations, in *Mathematics and computation, a contemporary view, vol. 3 of* Abel Symp., Springer, Berlin, 39–66.

Hou, T.Y. & Luo, G. (2013) On the finite-time blowup of a 1D model for the 3D incompressible Euler equations. arXiv:1311.2613.

Il'yashenko, J.S. (1992) Global analysis of the phase portrait for the Kuramoto–Sivashinsky equation. *J. Dynam. Differential Equations* **4**, no. 4, 585–615.

Kardar, M., Parisi, G., & Zhang, Y.-C. (1986) Dynamic scaling of growing interfaces. *Phys. Rev. Lett.* **56**, 889–892.

Kerr, R.M. (1993) Evidence for a singularity of the three-dimensional, incompressible Euler equations. *Phys. Fluids A* **5**, no. 7, 1725–1746.

Kinderlehrer, D. & Stampacchia, G. (1980) *An introduction to variational inequalities and their applications. Pure and Applied Mathematics* **88**. Academic Press Inc. [Harcourt Brace Jovanovich Publishers], New York.

Ladyzhenskaya, O.A., Solonnikov, V., & Ural'ceva, N. (1968) *Linear and quasilinear equations of parabolic type.* American Mathematical Society **23**, Providence, RI.

Lions, J.-L. & Magenes, E. *Non-homogeneous boundary value problems and applications.* Die Grundlehren der mathematischen Wissenschaften, Band 182. Springer-Verlag, New York, 1972.

Liu, X. (1991) Gevrey class regularity and approximate inertial manifolds for the Kuramoto–Sivashinsky equation. *Phys. D* **50**, no. 1, 135–151.

Luo, G. & Hou, T.Y. (2013) Potentially singular solutions of the 3D incompressible Euler equations. arXiv:1310.0497.

Luo, G. & Hou, T.Y. (2014a) Potentially singular solutions of the 3D axisymmetric Euler equations. PNAS, **111**, no. 36, 12968–12973.

Luo, G. & Hou, T.Y. (2014b) Toward the finite-time blowup of the 3D incompressible Euler equations: a numerical investigation. *SIAM MMS*, **12**, no. 4, 1722–1776.

Majda, A.J. & Bertozzi, A.L. (2002) *Vorticity and incompressible flow. Cambridge Texts in Applied Mathematics* **88**. Cambridge University Press, Cambridge.

Ohkitani, K. (2008) AMS MathSciNet review of the paper "On a generalization of the Constantin-Lax-Majda equation" by H. Okamoto, T. Sakajo, & M. Wunsch (see next entry).

Okamoto, H., Sakajo, T., & Wunsch, M. (2008) On a generalization of the Constantin–Lax–Majda equation. *Nonlinearity* **21**, no. 10, 2447–2461.

Pokhozhaev, S.I. (2008) On the blow-up of solutions of the Kuramoto–Sivashinsky equation. *Mat. Sb.* **199**, no. 9, 97–106.

Prodi, G. (1959) Un teorema di unicità per le equazioni di Navier–Stokes. *Ann. Mat. Pura Appl.* **48**, no. 4, 173–182.

Quittner, P. & Souplet, P. (2007) *Superlinear parabolic problems.* Birkhäuser Advanced Texts. Birkhäuser Verlag, Basel.

Robinson, J.C. (2001) *Infinite-dimensional dynamical systems.* Cambridge Texts in Applied Mathematics. Cambridge University Press, Cambridge.

Serrin, J. (1962) On the interior regularity of weak solutions of the Navier–Stokes equations. *Arch. Rational Mech. Anal.* **9**, 187–195.

Souplet, P. (2002) Gradient blow-up for multidimensional nonlinear parabolic equations with general boundary conditions. *Differential Integral Equations* **15**, no. 2, 237–256.

Takáč, P., Bollerman, P., Doelman, A., van Harten, A., & Titi, E.S. (1996) Analyticity of essentially bounded solutions to strongly parabolic semilinear systems. *SIAM Journal on Mathematical Analysis* **27**, 424–448.

Tadmor, E. (1986) The well-posedness of the Kuramoto–Sivashinsky equation. *SIAM J. Math. Anal.* **17**, no. 4, 884–893.

Temam, R. (1995) *Navier–Stokes equations and nonlinear functional analysis.* CBMS-NSF Regional Conference Series in Applied Mathematics **66**. Society for Industrial and Applied Mathematics (SIAM), Philadelphia, PA.

Temam, R. (1997) *Infinite-dimensional dynamical systems in mechanics and physics.* Applied Mathematical Sciences **68**. Springer-Verlag, New York.

Temam, R. (2001) *Navier–Stokes equations: theory and numerical analysis.* AMS Chelsea Publishing, Providence, RI. Reprint of the 1984 edition.

7

Parabolic Morrey spaces and mild solutions of the Navier–Stokes equations. An interesting answer through a silly method to a stupid question.

Pierre Gilles Lemarié–Rieusset

Laboratoire de Mathématiques et Modélisation d'Évry,
UMR CNRS 8071,
Université d'Évry. France.
`plemarie@univ-evry.fr`

Abstract

We present a theory of mild solutions for the Navier–Stokes equations in a (maximal) lattice Banach space.

7.1 The stupid question

Our question concerns the search for mild solutions of the Navier–Stokes problem. More precisely, let us consider the following Cauchy initial value problem for the Navier–Stokes equations on the whole space and with no external forces (and with viscosity taken equal to 1) :

$$\begin{cases} \partial_t \vec{u} = \Delta \vec{u} - (\vec{u}.\vec{\nabla})\vec{u} - \vec{\nabla}p \\ \vec{u}(0,x) = \vec{u}_0(x) \\ \operatorname{div} \vec{u} = 0 \end{cases} \tag{7.1}$$

When looking for a mild solution, one rewrites the problem as a fixed point problem of an integro–differential transform

$$\vec{u} = e^{t\Delta}\vec{u}_0 - B(\vec{u}, \vec{u}), \tag{7.2}$$

Published in *Recent Progress in the Theory of the Euler and Navier-Stokes Equations*, edited by James C. Robinson, José L. Rodrigo, Witold Sadowski, & Alejandro Vidal-López. ©Cambridge University Press 2016.

where the bilinear transform B is defined as

$$B(\vec{u}, \vec{v}) = \int_0^t e^{(t-s)\Delta} \mathbb{O}(\vec{u}(s,.) \otimes \vec{v}(s,.)) \, ds. \qquad (7.3)$$

$\mathbb{O}$ is the Oseen operator mapping matrix functions $F = (F_{ij})$ to vector functions $\vec{H} = (H_k)$ through the formula

$$H_k = \sum_{i,j} \mathbb{O}_{ijk} F_{ij} = \sum_{i,j} (\delta_{j,k}\partial_i - \frac{1}{\Delta}\partial_i\partial_j\partial_k) F_{i,j}. \qquad (7.4)$$

Mild solutions are then sought via Picard's iterative scheme: starting from $\vec{U}_0(t,x) = e^{t\Delta}\vec{u}_0$ and defining $\vec{U}_{n+1} = \vec{U}_0 - B(\vec{U}_n, \vec{U}_n)$, check whether the sequence $\vec{U}_n$ converges to a limit $\vec{u}$.

Our (stupid) question is then the following.

Question 1. Which is the largest space X such that taking $\|\vec{u}_0\|_X$ small enough implies that $\vec{U}_n$ converge to a global mild solution?

This is a 'stupid' question in the sense that the answer has been known for fifteen years (Koch & Tataru, 2001):

$$X = \mathrm{BMO}^{-1}.$$

Indeed, it is natural to ask that for our initial data $\vec{u}_0$, $e^{t\Delta}\vec{u}_0$ is locally square integrable (in order to give sense to (7.3)), i.e.

$$\int_0^1 \int_{B(0,1)} |e^{t\Delta}\vec{u}_0|^2 \, dt \, dx < +\infty.$$

Moreover, we should require this condition to be uniformly satisfied with respect to the invariance of the Navier–Stokes equations (i.e. if $\vec{u}_0$ is an admissible initial value, then so is $\lambda \vec{u}_0(\lambda(x-x_0))$; this requirement yields the condition

$$\sup_{t>0} \sup_{x_0 \in \mathbb{R}^3} \frac{1}{t^{3/2}} \int_0^t \int_{B(x_0,\sqrt{t})} |e^{s\Delta}\vec{u}_0(y)|^2 \, ds \, dy < +\infty.$$

This latter condition is in fact the characterization of distributions that belong to BMO^{-1}.

If we would like to alleviate the suspicion that we are dealing with an uninteresting problem, one may consider the same problem for the generalized Navier–Stokes problem where we replace the Laplacian operator by a fractional Laplacian operator,

$$\begin{cases} \partial_t \vec{u} = -(-\Delta)^{\alpha/2}\vec{u} - (\vec{u}.\vec{\nabla})\vec{u} - \vec{\nabla}p \\ \vec{u}(0,x) = \vec{u}_0(x) \\ \mathrm{div}\vec{u} = 0 \end{cases} \qquad (7.5)$$

where $1 < \alpha$.

The answer to the question then depends on the value of α:

- $\alpha = 2 :\ X = \mathrm{BMO}^{-1}$ [based on integration by parts]
- $1 < \alpha < 2 :\ X = \dot{B}_{\infty,\infty}^{1-\alpha}$ [no need to integrate by parts]
- $\alpha > 2 :$ unknown (at least, to me) [integration by parts does not work]

7.2 The silly method

Now, in order to try to provide an answer to Question 1, we are going to introduce a method that clearly cannot provide optimal answers (this is why I shall call it a silly method).

Let us recall that we have transformed the differential equation (7.1) into an (integro-)differential equation (7.2). Let $\mathbb{K}(t,x)$ be the integral kernel of the operator matrix $e^{t\Delta}\mathbb{O}$, so that the equation to be solved reads as

$$\vec{u} = e^{t\Delta}\vec{u}_0 + \int_0^t \int \mathbb{K}(t-s, x-y)(\vec{u}(s,y) \otimes \vec{u}(s,y))\,dy\,ds. \qquad (7.6)$$

As it is an integral equation, we want to use basic tools of integration such as Fatou's lemma, monotone convergence or dominated convergence. This is much easier when the integrand is nonnegative. Thus, we shall replace the equation (7.6) by a superequation

$$U(t,x) = |e^{t\Delta}\vec{u}_0| + \int_0^t \int |\mathbb{K}(t-s, x-y)|\, U^2(s,y))\,dy\,ds. \qquad (7.7)$$

While we gain in simplicity in the integral term, we definitely lose the main tool we have to control the solutions of the Navier–Stokes equations: we destroy any hope of using the dissipation expressed by the Leray energy inequality, which relies on cancellations in the nonlinear term.

Nonnegativity of the kernel is good, but it would be better if we also had symmetry. Thus, we shall use a further generalization of the equation, and consider

$$U(t,x) = 1_{t>0}|e^{t\Delta}\vec{u}_0| + \int_{s=-\infty}^{s=+\infty} \int |\mathbb{K}(|t-s|, x-y)|\, U^2(s,y)\,dy\,ds, \qquad (7.8)$$

where $1_{t>0}$ is 1 if $t > 0$ and zero otherwise.

The last bold step toward simplification will be to replace the kernel $\mathbb{K}$ by a simpler kernel. A well-known estimate states that we have

$$|\mathbb{K}(t,x)| \leq C_0 \frac{1}{t^2 + |x|^4} \tag{7.9}$$

for some positive constant C_0. So we shall consider the equation

$$U(t,x) = 1_{t>0}|e^{t\Delta}\vec{u}_0| + C_0 \int_{\mathbb{R}} \int \frac{1}{(t-s)^2 + |x-y|^4}\, U^2(s,y)\, \mathrm{d}y\, \mathrm{d}s. \tag{7.10}$$

More precisely, if $W_0(t,x)$, defined on $\mathbb{R} \times \mathbb{R}^3$, is such that the iterative sequence defined by induction from W_0 via

$$W_{n+1}(t,x) = W_0(t,x) + C_0 \int_{\mathbb{R}} \int \frac{1}{(t-s)^2 + |x-y|^4}\, W_n^2(s,y)\, \mathrm{d}y\, \mathrm{d}s \tag{7.11}$$

satisfies

$$\sup_{n\in\mathbb{N}} W_n(t,x) < \infty \quad \text{a.e. on } \mathbb{R} \times \mathbb{R}^3 \tag{7.12}$$

then we have the following consequences:

- $W(t,x) = \sup_{n\in\mathbb{N}} W_n(t,x)$ is a locally integrable function that satisfies

$$W(t,x) = W_0(t,x) + C_0 \int_{\mathbb{R}} \int \frac{1}{(t-s)^2 + |x-y|^4}\, W^2(s,y)\, \mathrm{d}y\, \mathrm{d}s. \tag{7.13}$$

- If we assume that $1_{t>0}|e^{t\Delta}\vec{u}_0| \leq W_0(t,x)$, $\vec{U}_0 = 1_{t>0}e^{t\Delta}\vec{u}_0$ and

$$\vec{U}_{n+1} = \vec{U}_0 - 1_{t>0}B(\vec{U}_n, \vec{U}_n)$$

then

$$|\vec{U}_{n+1}(t,x) - \vec{U}_n(t,x)| \leq W_{n+1}(t,x) - W_n(t,x) \tag{7.14}$$

so that

$$|\vec{U}_0(t,x)| + \sum_{n\in\mathbb{N}} |\vec{U}_{n+1}(t,x) - \vec{U}_n(t,x)| \leq W(t,x). \tag{7.15}$$

We therefore obtain a mild solution $\vec{u}$ of equation (7.2).

Thus, we are led to study the following questions.

Question 2. For which functions $W_0 \geq 0$ can we say that, for $\epsilon > 0$ small enough, we have a solution to the integral equation (7.13) for ϵW_0

$$W_\epsilon = \epsilon W_0 + \int_{\mathbb{R}} \int_{\mathbb{R}^3} \frac{C_0}{(t-s)^2 + |x-y|^4}\, W_\epsilon^2(s,y)\, \mathrm{d}s\, \mathrm{d}y?$$

Question 3. For which spaces of good initial values for the Navier–Stokes equations can we say that $W_0(t,x) = 1_{t>0}|e^{t,\Delta}\vec{u}_0|$ will fall into the class of functions required in Question 2?

More precisely, how much did we lose by changing Question 1 into Question 2?

7.3 Elliptic intermezzo

Before considering Question 2, we recall some basic facts involving integral equations with symmetric non-negative kernels (Lemarié–Rieusset, 2013). We thus look at the general integral equation

$$f(x) = f_0(x) + \int_X K(x,y)f^2(y)\,\mathrm{d}\mu(y) \qquad (7.16)$$

where μ is a non-negative σ-finite measure on a space X (i.e. $X = \cup_{n\in\mathbb{N}}Y_n$ with $\mu(Y_n) < +\infty$) and K is a positive measurable function on $X \times X$ with $K(x,y) > 0$ almost everywhere. We shall make a stronger assumption on K, namely that there exists a sequence X_n of measurable subsets of X such that $X = \cup_{n\in\mathbb{N}}X_n$ and

$$\int_{X_n}\int_{X_n}\frac{\mathrm{d}\mu(x)\,\mathrm{d}\mu(y)}{K(x,y)} < +\infty. \qquad (7.17)$$

Obviously, if f_0 is non-negative and f is an (a.e. finite) non-negative measurable solution of equation (7.16), then $0 \le f_0 \le f$ and

$$\int_X K(x,y)f^2(y)\,\mathrm{d}\mu(y) \le f(x) \quad \text{a.e.}$$

Conversely, if $0 \le f_0 < \frac{1}{4}\Omega$, with $\int_X K(x,y)\Omega^2(y)\,\mathrm{d}\mu(y) \le \Omega(x)$ a.e., then there exists an (a.e. finite) non-negative measurable solution of equation (7.16).

This gives the space in which to search for solutions of equation (7.16).

Proposition 7.3.1. *Let $\mathcal{E}_K$ be the space of measurable functions f on X such that there exist $\lambda \ge 0$ and a measurable non-negative function Ω such that $|f(x)| \le \lambda\Omega$ a.e. and $\int_X K(x,y)\Omega^2(y)\,\mathrm{d}\mu(y) \le \Omega(x)$ a.e. Then*

- $\mathcal{E}_K$ is a linear space and

$$\|f\|_K = \inf\{\lambda : \exists \, \Omega \geq 0 \text{ such that } |f| \leq \lambda\Omega$$

$$\text{and } \int_X K(x,y)\Omega^2(y) \, d\mu(y) \leq \Omega(x) \text{ a.e.}\}$$

  is a semi-norm on $\mathcal{E}_K$;
- $\|f\|_K = 0 \Leftrightarrow f = 0$ almost everywhere;
- the normed linear space E_K (obtained from $\mathcal{E}_K$ by quotienting with the relationship $f \sim g \Leftrightarrow f = g$ a.e.) is a Banach space; and
- if $f_0 \in \mathcal{E}_K$ is non-negative and satisfies $\|f_0\|_K < \frac{1}{4}$ then equation (7.16) has a non-negative solution $f \in \mathcal{E}_K$.

Our first example will be the elliptic nonlinear equation on $\mathbb{R}^d$ $(d \geq 3)$

$$-\Delta u = (-\Delta)^{1/2} u^2 - \Delta V.$$

This can be rewritten as

$$u = V + \mathcal{I}_1(u^2), \tag{7.18}$$

where the Riesz potential $\mathcal{I}_1$ is given by

$$\mathcal{I}_1 f(x) = \frac{1}{(-\Delta)^{1/2}} f(x) = \int_{\mathbb{R}^d} \frac{C_1}{|x-y|^{d-1}} f(y) \, dy.$$

The answer to Question 2 for equation (7.18) is well known.

Theorem 7.3.2 (Maz'ya & Verbitsky 1995). *If $V \geq 0$ then the following assertions are equivalent.*

(i) *For $\epsilon > 0$ small enough, we have a solution to the equation*

$$u_\epsilon = \epsilon V + \mathcal{I}_1(u_\epsilon^2).$$

(ii) *There exists a $C \geq 0$ such that V satisfies the inequality*

$$\int_{\mathbb{R}^d} V^2(x)(\mathcal{I}_1 f(x))^2 \, dx \leq C \int_{\mathbb{R}^d} f^2(x) \, dx \qquad \text{for all} \quad f \in L^2.$$

(iii) *V is a multiplier from the homogeneous Sobolev space $\dot{H}^1$ to L^2, i.e. there exists a $C \geq 0$ such that*

$$\int_{\mathbb{R}^d} V^2(x) f^2(x) \, dx \leq C \int_{\mathbb{R}^d} |\vec{\nabla} f(x)|^2 \, dx \qquad \text{for all} \quad f \in H^1.$$

Thus, we can see that the answer to Question 2 is far from being obvious. The maximal functional space in which to look for solutions is not a classical space, rather it is the space $\mathcal{V} = \mathcal{M}(\dot{H}^1 \mapsto L^2)$ of singular multipliers from $\dot{H}^1$ to L^2.

If we want to deal with some more amenable spaces, one can use the Fefferman–Phong inequality that relates the multiplier space to Morrey spaces. For $1 < p \le q < +\infty$, let us define the (homogeneous) Morrey space $\dot{M}^{p,q}$ in the following way: for $1 < p \le q < +\infty$

$$f \in \dot{M}^{p,q} \quad \text{if} \quad \sup_{R>0, x\in\mathbb{R}^d} R^{d(\frac{p}{q}-1)} \int_{|x-y|<R} |f(y)|^p \, dy < +\infty.$$

Theorem 7.3.3 (Fefferman–Phong 1983). *For $2 < p \le d$, we have*

$$\dot{M}^{p,d} \subset \mathcal{V} \subset \dot{M}^{2,d}.$$

Maz'ya and Verbitsky's theorem has been generalized to spaces of homogeneous type, i.e. spaces (X, δ, μ) such that

- for all $x, y \in X$, $\delta(x,y) \ge 0$;
- $\delta(x,y) = \delta(y,x)$;
- $\delta(x,y) = 0 \Leftrightarrow x = y$;
- there is a positive constant κ such that

$$\text{for all } x, y, z \in X, \delta(x,y) \le \kappa(\delta(x,z) + \delta(z,y)); \tag{7.19}$$

and
- there exist positive A, B and Q that satisfy

$$\text{for all } x \in X, \text{ for all } r > 0, \quad Ar^Q \le \int_{\delta(x,y)<r} d\mu(y) \le Br^Q. \tag{7.20}$$

Theorem 7.3.4 (Kalton & Verbitsky 1999). *Let (X, δ, μ) be a space of homogeneous type, set*

$$K_\alpha(x,y) = \frac{1}{\delta(x,y)^{Q-\alpha}} \tag{7.21}$$

(where $0 < \alpha < Q/2$) and let E_{K_α} be the associated Banach space (defined in Proposition 7.3.1). Let $\mathcal{I}_\alpha$ be the Riesz operator associated to K_α,

$$\mathcal{I}_\alpha f(x) = \int_X K_\alpha(x,y) f(y) \, d\mu(y). \tag{7.22}$$

We define two further linear spaces associated to K_α:

- *the Sobolev space W^α defined by*

$$g \in W^\alpha \Leftrightarrow \exists\, h \in L^2 \text{ such that } g = \mathcal{I}_\alpha h \qquad (7.23)$$

and

- *the multiplier space $\mathcal{V}^\alpha$ defined by*

$$f \in \mathcal{V}^\alpha \Leftrightarrow \|f\|_{\mathcal{V}^\alpha} = \left(\sup_{\|h\|_2 \le 1} \int_X |f(x)|^2 |\mathcal{I}_\alpha h(x)|^2 \, d\mu(x) \right)^{1/2} < +\infty$$

$$(7.24)$$

(so that pointwise multiplication by a function in $\mathcal{V}^\alpha$ is a bounded map from W^α to L^2).

Then we have (with equivalence of norms)

$$E_{K_\alpha} = \mathcal{V}^\alpha \qquad (7.25)$$

for $0 < \alpha < Q/2$.

7.4 Where we export our parabolic equations to the land of elliptic equations

Theorem 7.3.4 thus gives us the answer to our Question 2. Recall that we have transformed the "parabolic" Navier-Stokes equation (1.2)

$$\vec{u} = e^{t\Delta} \vec{u}_0 - B(\vec{u}, \vec{u})$$

into the "elliptic" equation (1.13)

$$W(t,x) = W_0(t,x) + C_0 \int_{\mathbb{R}} \int \frac{1}{(t-s)^2 + |x-y|^4} W^2(s,y) \, dy \, dx,$$

which we shall now interpret as

$$W = W_0 + \mathcal{J}_1(W^2), \qquad (7.26)$$

i.e. we shall consider $\mathbb{R} \times \mathbb{R}^3$ as a (parabolic) space of homogeneous type endowed with the quasi-norm $\rho(t,x) = (t^2 + |x|^4)^{1/4}$, in which case the homogeneous dimension Q is equal to $Q = 5$,

$$\iint_{B((t,x),R)} ds \, dy = cR^5,$$

and the Riesz potential has the form

$$\mathcal{J}_1 f(t,x) = \iint_{\mathbb{R} \times \mathbb{R}^3} \frac{C_0}{\rho(t-s, x-y)^{5-1}} f(s,y) \, ds \, dy.$$

134 P.G. Lemarié–Rieusset

The answer to Question 2 is then the following (Lemarié–Rieusset, 2013).

Theorem 7.4.1. *If $W_0 \geq 0$ then the following assertions are equivalent.*

(i) *For $\epsilon > 0$ small enough, we have a solution to the equation*
$$u_\epsilon = \epsilon W_0 + \mathcal{J}_1(u_\epsilon^2).$$

(ii) *There is a $C > 0$ such that W_0 satisfies the inequality*
$$\iint_{\mathbb{R}\times\mathbb{R}^3} W_0^2(t,x)(\mathcal{J}_1 f(t,x))^2 \, dt\, dx \leq C \iint_{\mathbb{R}\times\mathbb{R}^3} f^2(t,x)\, dt\, dx$$
for all $f \in L^2$.

(iii) *W_0 is a multiplier from the Sobolev space $\dot{H}_{t,x}^{\frac{1}{2},1}$ to L^2, i.e. there is a $C > 0$ such that*
$$\iint_{\mathbb{R}\times\mathbb{R}^3} W_0^2(t,x) f^2(t,x)\, dt\, dx \leq C \iint_{\mathbb{R}\times\mathbb{R}^3} |\Lambda f(t,x)|^2 \, dt\, dx$$
for all $f \in \dot{H}_{t,x}^{1/2,1}$, where $\Lambda = (-\partial_t^2)^{1/4} + (-\Delta_x)^{1/2}$.

7.5 Parabolic Morrey spaces and Triebel–Lizorkin-Morrey spaces, and other examples.

Obviously, we have a formal answer to our Question 3: the good space for the initial value should be the space of (divergence-free) vector fields such that $1_{t>0}|e^{t,\Delta}\vec{u}_0|$ is a multiplier from the Sobolev space $\dot{H}_{t,x}^{\frac{1}{2},1}$ to L^2. The problem is that this space is clearly not one of the classical spaces of functional analysis, so that, in a way, our answer is tautological: an initial value is good if it is a good initial value, whatever it actually means ...

If we want to get a better insight into what would be a good initial value, we may use a variant of the Fefferman–Phong inequality. We are thus going to compare our space of singular multipliers
$$\mathcal{W} = \mathcal{M}(\dot{H}_{t,x}^{\frac{1}{2},1} \mapsto L^2)$$
to parabolic Morrey spaces $\dot{M}^{p,q}(\mathbb{R}\times\mathbb{R}^3)$ $(1 < p \leq q < +\infty)$, where
$$f \in \dot{M}^{p,q}(\mathbb{R}\times\mathbb{R}^3) \Leftrightarrow \sup_{R>0,(t,x)\in\mathbb{R}\times\mathbb{R}^3} R^{5(\frac{p}{q}-1)} \int_{\rho(t-s,x-y)<R} |f(s,y)|^p \, ds\, dy < +\infty.$$

Theorem 7.5.1. *For $2 < p \leq 5$, we have*
$$\dot{M}^{p,5}(\mathbb{R}\times\mathbb{R}^3) \subset \mathcal{W} \subset \dot{M}^{2,5}(\mathbb{R}\times\mathbb{R}^3).$$

Then, a better (but partial) answer to our Question 3 would be to find a Banach space Y of measurable functions on $\mathbb{R} \times \mathbb{R}^3$ such that $Y \subset \mathcal{W}$ and to characterize the associated (maximal) Banach space X such that
$$\|1_{t>0}|e^{t,\Delta}\vec{u}_0|\|_Y \leq C\|\vec{u}_0\|_X$$
We may state some classical results in Navier–Stokes theory and check how we may easily show that they fit into our framework.

- The solutions of Fabes, Jones, & Rivière (1972) belong to the space $Y = L_t^p L_x^q$ with $\frac{2}{p} + \frac{3}{q} = 1$ and $3 \leq q < +\infty$; in this case we have $Y \subset \dot{M}^{\min(p,q),5}(\mathbb{R} \times \mathbb{R}^3) \subset \mathcal{W}$; the associated space is the Besov space $X = \dot{B}_{q,p}^{-\frac{2}{p}}$.
- Let us consider the limit case $p = +\infty$ and $q = 3$: the solutions of Kato (1984) belong to the space $Y = L_t^\infty L_x^3$; in this case we have $Y \subset \dot{M}^{3,5}(\mathbb{R} \times \mathbb{R}^3) \subset \mathcal{W}$ and the associated space is $X = L^3$.
- We may change the order of integration with respect to time and space. The space $Y = L_x^q L_t^p$ with $\frac{2}{p} + \frac{3}{q} = 1$ and $3 \leq q < +\infty$ will satisfy $Y \subset \dot{M}^{\min(p,q),5}(\mathbb{R} \times \mathbb{R}^3) \subset \mathcal{W}$ and the associated space is the Triebel–Lizorkin space $X = \dot{F}_{q,p}^{-\frac{2}{p}}$.
- In the limit case $p = +\infty$ and $q = 3$, we find the solutions of Calderón (1993) in the space $Y = L_x^3 L_t^\infty$; now $L_x^3 L_t^\infty \subset \dot{M}^{3,5}(\mathbb{R} \times \mathbb{R}^3) \subset \mathcal{W}$ and the associated space is $X = L^3$.

Further examples are discussed in Lemarié–Rieusset (2013, to appear). Of course, one should be interested in understanding as much as possible which space X corresponds to the (strange) space $Y = \mathcal{W}$. A close approach should be the investigation of the parabolic Morrey spaces. In this case, one recovers some already known spaces:

- We have $\mathcal{W} \subset \dot{M}^{2,5}(\mathbb{R} \times \mathbb{R}^3)$. It is worth noticing that the associated Banach space to $Y = \dot{M}^{2,5}(\mathbb{R} \times \mathbb{R}^3)$ is just the space $X = BMO^{-1}$ from Koch & Tataru (2001).
- On the other hand, we have $\dot{M}^{p,5}(\mathbb{R} \times \mathbb{R}^3) \subset \mathcal{W}$ when $2 < p \leq 5$. The associated Banach space to $Y = \dot{M}^{p,5}(\mathbb{R} \times \mathbb{R}^3)$ $(2 < p \leq 5)$ belongs to the scale of Triebel–Lizorkin–Morrey spaces studied by Sickel, Yang, & Yuan (2010): $X = \dot{F}_{p,p}^{-\frac{2}{p},\frac{1}{p}-\frac{1}{q}}$ with $\frac{2}{p} + \frac{3}{q} = 1$. This may be the first "natural" setting where these spaces appear.

References

Calderón, C. (1993) Initial values of Navier–Stokes equations. *Proc. Amer. Math. Soc.* **117**, 761–766.

Fabes, E., Jones, B.F., & Rivière, N. (1972) The initial value problem for the Navier–Stokes equations with data in L^p. *Arch. Rat. Mech. Anal.* **45**, 222–240.

Fefferman, C. (1983) The uncertainty principle. *Bull. Amer. Math. Soc.* **9**, 129–206.

Kalton, N. & Verbitsky, I. (1999) Nonlinear equations and weighted norm inequalities. *Trans. Amer. Math. Soc.* **351**, 3441–3497.

Kato, T. (1984) Strong L^p solutions of the Navier–Stokes equations in $\mathbb{R}^m$ with applications to weak solutions. *Math. Zeit.* **187** 471–480.

Koch, H. & Tataru, D. (2001) Well-posedness for the Navier–Stokes equations. *Advances in Math.* **157**, 22–35.

Lemarié–Rieusset, P.G. (2013) Sobolev multipliers, maximal functions and parabolic equations with a quadratic nonlinearity. Preprint, Univ. Evry.

Lemarié–Rieusset, P.G. (2015) *The Navier–Stokes equations in the XXIst century.* CRC Press, to appear.

Maz'ya, V. & Verbitsky, I. (1995) Capacitary inequalities for fractional integrals, with applications to partial differential equations and Sobolev multipliers. *Ark. Mat.* **33**, 81–115.

Sickel, W., Yang, D., & Yuan, W. (2010) *Morrey and Campanato meet Besov, Lizorkin and Triebel.* Lecture Notes in Math. **2005**, Springer.

8

Well-posedness for the diffusive 3D Burgers equations with initial data in $H^{1/2}$

Benjamin C. Pooley

Mathematics Institute, University of Warwick,
Coventry, CV4 7AL. UK.
b.c.pooley@warwick.ac.uk

James C. Robinson

Mathematics Institute, University of Warwick,
Coventry, CV4 7AL. UK.
j.c.robinson@warwick.ac.uk

Abstract

In this note we discuss the diffusive, vector-valued Burgers equations in a three-dimensional domain with periodic boundary conditions. We prove that given initial data in $H^{1/2}$ these equations admit a unique global solution that becomes classical immediately after the initial time. To prove local existence, we follow as closely as possible an argument giving local existence for the Navier–Stokes equations. The existence of global classical solutions is then a consequence of the maximum principle for the Burgers equations due to Kiselev & Ladyzhenskaya (1957).

In several places we encounter difficulties that are not present in the corresponding analysis of the Navier–Stokes equations. These are essentially due to the absence of any of the cancellations afforded by incompressibility, and the lack of conservation of mass. Indeed, standard means of obtaining estimates in L^2 fail and we are forced to start with more regular data. Furthermore, we must control the total momentum and carefully check how it impacts on various standard estimates.

Published in *Recent Progress in the Theory of the Euler and Navier-Stokes Equations*, edited by James C. Robinson, José L. Rodrigo, Witold Sadowski, & Alejandro Vidal-López. ©Cambridge University Press 2016.

8.1 Introduction

We consider the three-dimensional, vector-valued diffusive Burgers equations. The equations, for a fixed viscosity $\nu > 0$ and initial data u_0, are

$$u_t + (u \cdot \nabla)u - \nu \Delta u = 0, \tag{8.1}$$

$$u(0) = u_0. \tag{8.2}$$

Working on the torus $\mathbb{T}^3 = \mathbb{R}^3/2\pi\mathbb{Z}^3$, we will investigate the existence and uniqueness of solutions u to (8.1). Using the rescaling $\tilde{u}(x,t) := \nu u(x,\nu t)$, it suffices to prove well-posedness in the case $\nu = 1$.

This system is well known and is often considered to be 'well understood'. However we have not found a self-contained account of its well–posedness in the literature, although for very regular data (with two Hölder continuous derivatives, for example) existence and uniqueness can be deduced from standard results about quasi–linear systems. We are particularly interested in an analysis parallel to the familiar treatment of the Navier–Stokes equations, which motivates our choice of function spaces here.

It is interesting to note that we find some essential difficulties in treating this system which do not occur when incompressibility is enforced, i.e. for the Navier–Stokes equations. These prevent us from making the usual estimates that would give existence of (L^2-valued) weak solutions. We also find that taking initial data with zero average is not sufficient to ensure that the solution has zero average for positive times. This necessitates estimating the momentum and checking carefully that the methods applicable to the Navier–Stokes equations have a suitable analogue.

We begin with some brief comments on several relevant methods from the literature to motivate our discussion here.

A maximum principle for solutions of the Burgers equations was first proved by Kiselev & Ladyzhenskaya (1957). A simplified version of this result with zero forcing plays a key role in our argument, so we reproduce the proof here.

Lemma 8.1.1. *If u is a classical solution of the Burgers equations (8.1) on a time interval $[a, b]$ then*

$$\sup_{t \in [a,b]} \|u(t)\|_{L^\infty} \leq \|u(a)\|_{L^\infty}. \tag{8.3}$$

Proof. Fix $\alpha > 0$ and let $v(t,x) := e^{-\alpha t}u(x,t)$ for all $x \in \mathbb{T}^3$. Then $|v|^2$

satisfies the equation

$$\frac{\partial}{\partial t}|v|^2 + 2\alpha|v|^2 + u \cdot \nabla|v|^2 - 2v \cdot \Delta v = 0. \tag{8.4}$$

Since $2v \cdot \Delta v = \Delta|v|^2 - 2|\nabla v|^2$ we see that if $|v|^2$ has a local maximum at $(x, t) \in (a, b] \times \mathbb{T}^3$ then the left-hand side of (8.4) is positive unless $|v(x, t)| = 0$. Hence

$$\|u(t)\|_{L^\infty} \le e^{\alpha t}\|u(a)\|_{L^\infty}.$$

Now (8.3) follows because $\alpha > 0$ was arbitrary. $\qquad\square$

In the discussion of well-posedness for (8.1) in Kiselev & Ladyzhenskaya (1957) the maximum principle is used with approximations obtained by considering discrete times and replacing the time derivatives by difference quotients. Unfortunately one of the steps there is incorrect. In the MathSciNet review, R. Finn relates a comment by L. Nirenberg that there is a flaw in the compactness argument given on p. 675. This error appears to be fatal.

Another well known approach comes by analogy with the Burgers equations in one dimension, namely the Cole–Hopf transformation, which gives analytic solutions by reducing the problem to solving a heat equation. Unfortunately this can only give gradient solutions, and since we wish to draw comparisons with the classical equations of fluid mechanics this is a significant drawback.

There is a theorem in the book of Ladyzhenskaya, Solonnikov, & Ural'ceva (1968) (Chapter VII, Theorem 7.1) giving local well–posedness for a certain class of quasi-linear parabolic problems that includes (8.1). In that theorem the data and solutions are taken to have spatial Hölder regularity[1] at least $C^{2,\alpha}$ for some $\alpha \in (0, 1)$. It is likely that a consequence is global well–posedness in these spaces, but this is not stated. A brief sketch of the proof is given, but it is quite different from any familiar method used for the Navier–Stokes equations. Moreover and there is also no discussion of solutions gaining regularity that we will demonstrate (see Lemma 8.2.5).

To simplify several of the estimates proved later, we define for $s \ge 0$ the operator Λ^s acting on $H^s(\mathbb{T}^3)$ as follows. Let $f \in H^s(\mathbb{T}^3)$ have the

---

[1]The spaces in which solutions are found are actually defined by the existence and Hölder continuity (with certain exponent) of the mixed derivatives $D_t^r D_x^s$ for $2r + s < 2 + \alpha$, where D_x^s is any spatial derivative of order s.

Fourier series

$$f(x) = \sum_{k \in \mathbb{Z}^3} \widehat{f}_k e^{ik \cdot x} \in H^s(\mathbb{T}^3),$$

then we define

$$\Lambda^s f(x) := \sum_{k \in \mathbb{Z}^3} |k|^s \widehat{f}_k e^{ik \cdot x} \in L^2(\mathbb{T}^3).$$

Moreover we denote by $\| \cdot \|_s$ the seminorm $\|\Lambda^s \cdot \|_{L^2}$. This is of course compatible with the definition of the Sobolev norm; $\| \cdot \|_{H^s}$ is equivalent to $\| \cdot \|_{L^2} + \| \cdot \|_s$. Note that in the Fourier setting it is more usual to define an equivalent norm on H^s by

$$\|f\| = \left(\sum_{k \in \mathbb{Z}^3} (1 + |k|^{2s}) |\widehat{f}_k|^2 \right)^{1/2},$$

but here we will usually consider $\|f\|_{L^2}$ and $\|f\|_s$ separately, when estimating $\|f\|_{H^s}$. We will also make use of the fact that $\|f\|_s \le \|f\|_t$ if $0 < s \le t$ and that $\Lambda^2 = (-\Delta)$,

We call $u \in C^0([0, T]; H^{1/2}) \cap L^2(0, T; H^{3/2})$ with $u_t \in L^2(0, T; H^{-1/2})$ a *strong solution* of (8.1) if, for any $\phi \in C^\infty(\mathbb{T}^3)$

$$\langle u_t, \phi \rangle + ((u \cdot \nabla)u, \phi)_{L^2} + (\nabla u, \nabla \phi)_{L^2} = 0 \qquad (8.5)$$

for almost all $t \in [0, T]$. Here $\langle \cdot, \cdot \rangle$ denotes the duality pairing of $H^{-1/2}(\mathbb{T}^3)$ with $H^{1/2}(\mathbb{T}^3)$. We consider the attainment of the initial data $u_0 \in H^{1/2}$ in the sense of continuity into $H^{1/2}$.

We have chosen to use the term *strong solution* here, even though (in the classical treatment of the Navier–Stokes equations) this usually refers to solutions in $L^\infty(0, T; H^1) \cap L^2(0, T; H^2)$. Indeed, we shall see that the solutions we find become classical, in a similar way to local strong solutions of the Navier–Stokes equations (see Robinson, Rodrigo, & Sadowski, 2016).

The reason for considering well-posedness in $H^{1/2}$ is that, as for the Navier–Stokes equations, if u is a solution to the Burgers equations on $\mathbb{R}^3$ then, for $\lambda > 0$, so is

$$u_\lambda := \lambda u(\lambda^2 t, \lambda x)$$

and in three dimensions the seminorm $\| \cdot \|_{1/2}$ is invariant under this scaling. Therefore we would ideally consider solutions in the homogeneous space $\dot{H}^{1/2}$; however, as we will see, the zero-average property is not necessarily preserved in the solution. Fortunately we will also see that it is natural to control the 'creation of momentum' by $\int_0^t \|u(s)\|_{1/2}\,\mathrm{d}s$ and we will check carefully that the relevant techniques from the analysis of the Navier–Stokes equations in $\dot{H}^{1/2}$ can be adapted.

We will prove the following theorem.

Theorem 8.1.2. *Given $u_0 \in H^{1/2}$, there exists a unique global strong solution $u \in C^0([0,\infty); H^{1/2}) \cap L^2(0,\infty; H^{3/2})$. Moreover, except at the initial time, $u \in C^1((0,\infty); C^0) \cap C^0((0,\infty); C^2)$ and is a classical solution.*

We will prove this using Galerkin approximations to find unique local strong solutions and then, by bootstrapping, prove that the solution has enough regularity to rule out a blowup and deduce global existence using Lemma 8.1.1.

The well known arguments giving global existence of weak solutions to the Navier–Stokes equations (i.e. solutions with initial data in L^2 and regularity $L^\infty(0,T; L^2) \cap L^2(0,T; H^1))$ rely on the anti-symmetry

$$((u \cdot \nabla)v, w)_{L^2} = -((u \cdot \nabla)w, v)_{L^2}$$

that is a consequence of incompressibility of u. This is not something we can make use of with the Burgers equations. We might instead try to find weak solutions using the maximum principle and the following estimate that holds for smooth solutions

$$\frac{\mathrm{d}}{\mathrm{d}t}\|u\|_{L^2}^2 + \|\nabla u\|_{L^2}^2 \leq \|u\|_{L^2}^2 \|u\|_{L^\infty}^2. \tag{8.6}$$

Making rigorous use of this would require a maximum principle to hold for the Galerkin approximations, but the proof of Lemma 8.1.1 does not work with a projection applied to the nonlinear term.

To avoid these difficulties we will start with more regular initial data (in $H^{1/2}$) and find classical solutions before making use of Lemma 8.1.1.

We separate the difficulties encountered in the proof of Theorem 8.1.2 into two sections. In Section 8.2 we prove global well-posedness for data $u_0 \in H^1$. Here we use some standard *a priori* estimates to find local strong solutions. We then bootstrap to show that the solution is classical after the initial time. This allows us to apply Lemma 8.1.1, from which we derive better H^1 estimates that imply global existence.

In Section 8.3 we prove Theorem 8.1.2 using techniques from Marín-Rubio, Robinson, & Sadowski (2013) to find a unique local solution for initial data $u_0 \in H^{1/2}$. This solution instantly becomes classical, and hence global, by the results in Section 8.2.

8.2 Solutions in H^1

We will use the method of Galerkin approximations. First we introduce some notation. For $n \in \mathbb{N}$ let P_n denote the projection onto the Fourier modes of order up to n, that is

$$P_n \left(\sum_{k \in \mathbb{Z}^3} \widehat{u}_k e^{\mathrm{i}x \cdot k} \right) = \sum_{|k| \leq n} \widehat{u}_k e^{\mathrm{i}x \cdot k}.$$

Let $u_n = P_n u_n$ be the solution to

$$\frac{\partial u_n}{\partial t} + P_n[(u_n \cdot \nabla)u_n] - \Delta u_n = 0, \tag{8.7}$$

with

$$u_n(0) = P_n u_0. \tag{8.8}$$

For some maximal $T_n > 0$ there exists a solution $u_n \in C^{\infty}([0, T_n) \times \mathbb{T}^3)$ to this finite-dimensional locally-Lipschitz system of ODEs.

As noted in the introduction, one of the interesting issues we encounter in this analysis of the Burgers equations is that we cannot guarantee that the solution has the zero-average property even if the initial data does. However we do have the following estimate to control the potential 'creation of momentum'.

Lemma 8.2.1. *Let u, v be solutions of (8.7), with initial data u_0 and v_0 respectively. If $w = u - v$ and $w_0 = u_0 - v_0$ then*

$$\left| \int_{\mathbb{T}^3} w(x, t) - w_0(x) \, \mathrm{d}x \right| \leq 8\pi^3 \int_0^t \|w(s)\|_{1/2}(\|u(s)\|_{1/2} + \|v(s)\|_{1/2}) \, \mathrm{d}s. \tag{8.9}$$

In particular, taking $v \equiv 0$ yields

$$\left| \int_{\mathbb{T}^3} u(x, t) \, \mathrm{d}x \right| \leq 8\pi^3 \int_0^t \|u(s)\|_{1/2}^2 \, \mathrm{d}s + \left| \int_{\mathbb{T}^3} u_0(x) \, \mathrm{d}x \right|. \tag{8.10}$$

Proof. For $k \in \mathbb{Z}^3$ denote the kth Fourier coefficients of u, v and w by $\widehat{u}_k$, $\widehat{v}_k$ and $\widehat{w}_k$ respectively. Considering the form of the equations satisfied by u and v, we have

$$\frac{\mathrm{d}}{\mathrm{d}t} \int_{\mathbb{T}^3} w(x, t) \, \mathrm{d}x = - \int_{\mathbb{T}^3} (u \cdot \nabla)w + (w \cdot \nabla)v \, \mathrm{d}x$$

$$= -8\pi^3 \mathrm{i} \sum_{k \in \mathbb{Z}^3} \left(\overline{\widehat{u}_k(t)} \cdot k \right) \widehat{w}_k(t) + \left(\overline{\widehat{w}_k(t)} \cdot k \right) \widehat{v}_k(t),$$

Hence

$$\left| \frac{\mathrm{d}}{\mathrm{d}t} \int_{\mathbb{T}^3} w(x,t)\,\mathrm{d}x \right| \leq 8\pi^3 \sum_{k \in \mathbb{Z}^3} |\widehat{w}_k| |k| (|\widehat{u}_k| + |\widehat{v}_k|)$$

$$\leq 8\pi^3 \|w(t)\|_{1/2} (\|u(t)\|_{1/2} + \|v(t)\|_{1/2}),$$

then (8.9) follows after integrating with respect to t. $\qquad\square$

We will use this lemma to control the failure of equivalence of the norm $\|\cdot\|_{H^s}$ and the seminorm $\|\cdot\|_s$ for solutions of (8.7) as follows:

$$\|u_n(t)\|_s \leq \|u_n(t)\|_{H^s} \leq c\|u_n(t)\|_s + c \int_0^t \|u_n\|_{1/2}^2\,\mathrm{d}s + c\|u_0\|_{L^1}, \quad (8.11)$$

for some $c > 0$ depending only on s. Here we have used the fact that $\int_{\mathbb{T}^3} P_n u_0 = \int_{\mathbb{T}^3} u_0$. Note that we will occasionally use the equivalence of the seminorms $\|\cdot\|_s$ and $\|\cdot\|_{\dot{H}^s}$ when applicable. In particular for estimating the derivatives of sufficiently regular functions, e.g. $\|\nabla u_n\|_{L^6} \leq c\|u_n\|_2$.

We will prove the following special case of Theorem 8.1.2. The proofs of some estimates will only be sketched, if they are standard or when similar arguments are made in detail in Section 8.3.

Theorem 8.2.2. *Given $u_0 \in H^1$, there exists a unique global strong solution $u \in C^0([0,\infty); H^1) \cap L^2(0,\infty; H^2)$. Moreover, except at the initial time, $u \in C^1((0,\infty); C^0) \cap C^0((0,\infty); C^2)$ is a classical solution.*

We first need a lower bound on the existence times for the Galerkin systems (8.7) that is uniform, i.e. independent of n. For this we integrate the L^2 inner product of (8.7) with $\Lambda^2 u_n$. Using the inequalities of Hölder and Young to control the nonlinear term, as we would for the Navier–Stokes equations, we obtain

$$\|u_n(t)\|_1^2 + \int_0^t \|u_n(s)\|_2^2\,\mathrm{d}s \leq \|u_n(0)\|_1^2 + c \int_0^t \|u_n(s)\|_{L^6}^4 \|u_n(s)\|_1^2\,\mathrm{d}s$$

for some $c > 0$. Now by the embedding $H^1 \hookrightarrow L^6$ and Lemma 8.2.1

$$\|u_n(t)\|_1^2 + \int_0^t \|u_n(s)\|_2^2 \, \mathrm{d}s \leq \|u_n(0)\|_1^2 + c \int_0^t \|u_n(s)\|_1^6 \, \mathrm{d}s$$

$$+ c \left(\int_0^t \|u_n(s)\|_{1/2}^2 \, \mathrm{d}s + \|u_0\|_{L^1} \right)^4 \int_0^t \|u_n(s)\|_1^2 \, \mathrm{d}s$$

$$\leq \|u_n(0)\|_1^2 + c \int_0^t \|u_n(s)\|_1^6 + c \int_0^t t^4 \|u_n(s)\|_1^{10} + \|u_0\|_{L^1}^4 \|u_n(s)\|_1^2 \, \mathrm{d}s,$$

$$(8.12)$$

for some $c > 0$. The last step made use of the fact that $\|u_n\|_{1/2} \leq \|u_n\|_1$ and the Hölder inequality

$$\left(\int_0^t f(s) \, \mathrm{d}s \right)^5 \leq t^4 \int_0^t |f(s)|^5 \, \mathrm{d}s.$$

Let us now impose the upper bound $t \leq 1$. Applying Young's inequality to the $\|u_n\|_1^6$ and $\|u_n\|_1^2$ terms of the last line of (8.12) gives

$$\|u_n(t)\|_1^2 \leq \|u_n(0)\|_1^2 + c \int_0^t \alpha^5 \|u_n(s)\|_1^{10} + \beta^5 \, \mathrm{d}s,$$

where $\alpha := (4 + \|u_0\|_{L^1}^4)^{1/5}$ and $\beta := (2 + 4\|u_0\|_{L^1}^4)^{1/5}$. This gives an integral inequality of the form

$$f(t) \leq f(0) + \int_0^t (af(s) + b)^5 \, \mathrm{d}s.$$

By solving this inequality we obtain

$$\|u_n(t)\|_1^2 \leq \frac{\alpha \|u_0\|_1^2 + \beta}{\alpha \left(1 - 4\alpha t (\alpha \|u_0\|_1^2 + \beta)^4 \right)^{1/4}} - \frac{\beta}{\alpha}. \qquad (8.13)$$

Using Lemma 8.2.1 this estimate rules out a blowup of u_n in H^1 before the time

$$T^* := \frac{1}{4\alpha (\alpha \|u_0\|_1^2 + \beta)^4}. \qquad (8.14)$$

It follows that there exists $T > 0$, we can for example take $T = T^*/2$, such that $T_n \geq T$ for all N. At this point one might optimize the existence time T by choosing a different upper bound where we assumed $t \leq 1$, but this is not necessary in what follows.

From (8.12) and (8.13) we now have uniform bounds on u_n in the spaces $L^\infty(0, T; H^1)$ and $L^2(0, T; H^2)$. By (8.7) and a standard argument, we also obtain a uniform bound on the time derivative $\partial_t u_n \in$

$L^2(0, T; L^2)$. Therefore, by the Aubin-Lions theorem, we may assume (after passing to a subsequence) that there exists $u \in C^0([0, T]; H^1)$ such that $u_n \to u$ in $L^p(0, T; L^2)$ for all $p \in (1, \infty)$. A standard method shows that the limit u is a strong solution. For the details of this type of argument, see, for example, Constantin & Foias (1988), Evans (2010), Galdi (2000) or Robinson (2001).

It can also be shown that this solution u is a unique. We omit the proof of this here but we will see, in Section 8.3, that uniqueness requires even less regularity. Subject to this omission we have so far proved the following.

Lemma 8.2.3. *Given $u_0 \in H^1$ there exists $T > 0$ (given by (8.14)) such that the Burgers equations admit a unique strong solution u on $[0, T]$ with initial data u_0. Moreover $u \in C^0([0, T]; H^1) \cap L^2(0, T; H^2)$.*

It follows that if $u \in L^\infty(0, T; H^{1/2}) \cap L^2(0, T; H^{3/2})$ is a strong solution of the Burgers equations then $u \in C^0([\varepsilon, T]; H^1) \cap L^2(\varepsilon, T; H^2)$ for any $\varepsilon \in (0, T)$. Therefore the following corollary is an easy consequence of (8.14).

Corollary 8.2.4. *If $u \in L^\infty(0, T; H^{1/2}) \cap L^2(0, T; H^{3/2})$ is a local strong solution of (8.5) such that $T \in (0, \infty)$ is the maximum existence time (i.e. no strong solution exists on $[0, T + \varepsilon]$ for any $\varepsilon > 0$), then* ess $\sup_{(0,T)} \|u(t)\|_{H^1} = \infty$.

In order to prove Theorem 8.2.2, it remains to show that u is, in fact, a global classical solution after the initial time. We will use a bootstrapping argument to obtain local classical solutions, followed by the maximum principle that will allow us to apply Corollary 8.2.4 to show that the solution can be extended for an arbitrary length of time.

The bootstrapping is carried out with the following lemma which is actually stronger than we will need. We will omit the proof as it is essentially the same as standard results about strong solutions of the Navier–Stokes equations that can be found in Constantin & Foias (1988) and Robinson (2006), for example.

Lemma 8.2.5. *If the Galerkin approximations u_n are uniformly bounded in $L^2(\varepsilon, T; H^{s+1}) \cap L^\infty(\varepsilon, T; H^s)$ for $s > 1/2$ and some $\varepsilon \geq 0$, then they are also bounded uniformly in $L^2(\varepsilon', T; H^{s+2}) \cap L^\infty(\varepsilon', T; H^{s+1})$ for any $\varepsilon' \in (\varepsilon, T)$.*

The uniform bounds on u_n we have proved are sufficient to apply this lemma. In particular by applying it five times we have that for any

$\varepsilon \in (0,T)$, $(u_n)_{n=1}^{\infty}$ is a uniformly bounded sequence in $L^{\infty}(\varepsilon, T; H^6)$. In this case we have the following estimates on the time derivatives of u_n:

$$\sup_{t\in(\varepsilon,T)}\left\|\frac{\partial u_n}{\partial t}\right\|_{H^4} \leq \sup_{t\in(\varepsilon,T)}\left(\|u_n(t)\|_{H^4}\|u_n(t)\|_{H^5} + \|u_n(t)\|_{H^6}\right),$$

and

$$\sup_{t\in(\varepsilon,T)}\left\|\frac{\partial^2 u_n}{\partial t^2}(t)\right\|_{H^2} \leq \sup_{t\in(\varepsilon,T)}\left(\left\|\frac{\partial u_n}{\partial t}(t)\right\|_{H^2}\|u_n(t)\|_{H^3}\right)$$

$$+ \sup_{t\in(\varepsilon,T)}\left(\|u_n(t)\|_{H^2}\left\|\frac{\partial u_n}{\partial t}(t)\right\|_{H^3} + \left\|\frac{\partial u_n}{\partial t}(t)\right\|_{H^4}\right)$$

It follows that (u_n) is a bounded sequence in $H^1(\varepsilon, T; H^4)\cap H^2(\varepsilon, T; H^2)$. This regularity passes to the limit i.e. $u \in H^1(\varepsilon, T; H^4) \cap H^2(\varepsilon, T; H^2)$ and hence $u \in C^0([\varepsilon, T]; C^2) \cap C^1([\varepsilon, T]; C^0)$. This is enough regularity to conclude that u is a local classical solution of the Burgers equations. Note that time regularity on these closed intervals follows by considering larger open intervals.

To show that u can be extended to a global solution we now use the maximum principle from Lemma 8.1.1. Taking $\varepsilon > 0$ as the initial time of the classical solution, as above, we have the following estimate for $t \in [\varepsilon, T]$:

$$\frac{\mathrm{d}}{\mathrm{d}t}\|u\|_1^2 \leq 2|((u \cdot \nabla)u, \Lambda^2 u)_{L^2}| - 2\|u\|_2^2 \leq \|u\|_{L^\infty}^2\|u\|_1^2.$$

Therefore

$$\sup_{t\in[\varepsilon,T]}\|u(t)\|_1 \leq \|u(\varepsilon)\|_1 e^{t\|u(\varepsilon)\|_{L^\infty}^2/2}.$$

This rules out the blowup of u in the H^1 norm as $t \to T$, hence by Corollary 8.2.4 the solution can be extended over $[\varepsilon, \infty)$, as required.

This completes the proof of Theorem 8.2.2, subject to a proof of uniqueness which can be found in the next section.

8.3 Proof of Theorem 8.1.2

We now set about proving Theorem 8.1.2. The argument will follow the same pattern as the previous section. That is, we will prove that for initial data in $H^{1/2}$ there exists T, independent of n, such that the Galerkin systems have solutions on an interval $[0, T]$. We then deduce the existence and uniqueness of a local strong solution $u \in L^2(0, T; H^{3/2}) \cap C^0([0, T]; H^{1/2})$ of the Burgers equations. This is regular enough that global solutions can be obtained by appealing to the case of H^1 data.

As in the previous section we denote by $u_n \in C^\infty([0, T_n] \times \mathbb{T}^3)$ the unique solution to the Galerkin system (8.7) with maximal existence time T_n. We allow the case $T_n = \infty$ but note that if $T_n < \infty$ then we necessarily have $\|u_n(t)\|_{L^2} \to \infty$ as $t \nearrow T_n$.

Following Marín-Rubio et al. (2013) (see also Chemin et al. (2006), Calderón (1990) and Fabes, Jones & Rivière (1972)) we split (8.7) into a heat part, and a nonlinear part with zero initial data. Let v be the periodic solution of the heat equation with initial data u_0, then $v_n := P_n v$ satisfies

$$\frac{\partial}{\partial t} v_n + \Delta v_n = 0, \ v_n(0) = P_n u_0.$$

Let $w_n := u_n - v_n$, then w_n satisfies

$$\frac{\partial}{\partial t} w_n + P_n[(u_n \cdot \nabla)u_n] - \Delta w_n = 0, \ w_n(0) = 0. \tag{8.15}$$

For v_n and $t \in [0, T_n)$ we have the estimate

$$\sup_{s \in [0,t]} \|v_n(s)\|_{H^{1/2}}^2 + 2 \int_0^t \|v_n(s)\|_{H^{3/2}}^2 \, \mathrm{d}s \le \|P_n u_0\|_{H^{1/2}}^2. \tag{8.16}$$

Integrating (8.15) against $\Lambda^1 w_n$, gives

$$\|w_n(t)\|_{1/2}^2 + 2 \int_0^t \|w_n(s)\|_{3/2}^2 \, \mathrm{d}s$$

$$\le \int_0^t \|u_n(s)\|_{L^6} \|\nabla u_n(s)\|_{L^2} \|\Lambda^1 w_n(s)\|_{L^3} \, \mathrm{d}s \tag{8.17}$$

$$\le c_1 \int_0^t \|u_n(s)\|_{H^1} \|u_n(s)\|_1 \|w_n(s)\|_{3/2} \, \mathrm{d}s =: I_0.$$

For some $c_1 > 0$. Now by Lemma 8.2.1 and the definition of w_n,

$$\|u_n(t)\|_{H^1} \|u_n(t)\|_1 \le c_2 \left(\|u_n(t)\|_1 + \int_0^t \|u_n(s)\|_{1/2}^2 \, \mathrm{d}s + \|u_0\|_{L^1} \right) \|u_n(t)\|_1$$

$$\le 2c_2 (\|v_n(t)\|_1^2 + \|w_n(t)\|_1^2)$$

$$+ c_2 (\|v_n(t)\|_1 + \|w_n(t)\|_1) \left(\int_0^t \|u_n(s)\|_{1/2}^2 \, \mathrm{d}s + \|u_0\|_{L^1} \right) =: I_1 + I_2$$

for some $c_2 > 0$. To estimate $I_1 \times \|w_n\|_{3/2}$ we apply Young's inequality,

$$\|w_n(t)\|_{3/2}(\|v_n(t)\|_1^2 + \|w_n(t)\|_1^2)$$

$$\leq \frac{1}{4c_1c_2}\|w_n(t)\|_{3/2}^2 + c\|v_n(t)\|_1^4 + \|w_n(t)\|_{3/2}\|w_n(t)\|_1^2,$$

for some $c > 0$. Also by several applications of Young's inequality, we estimate $I_2 \times \|w_n\|_{3/2}$ as follows:

$$\|w_n(t)\|_{3/2}(\|v_n(t)\|_1 + \|w_n(t)\|_1)\left(\int_0^t \|u_n(s)\|_{1/2}^2\,\mathrm{d}s + \|u_0\|_{L^1}\right)$$

$$\leq \frac{1}{2}\|w_n(t)\|_{3/2}\left(\|v_n(t)\|_1^2 + \|w_n(t)\|_1^2\right)$$

$$+ \|w_n(t)\|_{3/2}\left(\int_0^t \|u_n(s)\|_{1/2}^2\,\mathrm{d}s + \|u_0\|_{L^1}\right)^2$$

$$\leq \frac{1}{2c_1c_2}\|w_n(t)\|_{3/2}^2 + c\|v_n(t)\|_1^4 + \frac{1}{2}\|w_n(t)\|_{3/2}\|w_n(t)\|_1^2$$

$$+ c\left(\int_0^t \|u_n(s)\|_{1/2}^2\,\mathrm{d}s + \|u_0\|_{L^1}\right)^4$$

for some $c > 0$. To control the $\|w_n\|_{3/2}\|w_n\|_1^2$ terms in the last two estimates we use the interpolation

$$\int_0^t \|w_n(s)\|_{3/2}\|w_n(s)\|_1^2\,\mathrm{d}s \leq \int_0^t \|w_n(s)\|_{3/2}^2\|w_n(s)\|_{1/2}\,\mathrm{d}s$$

$$\leq \frac{1}{5c_1c_2}\sup_{s\in[0,t]}\|w_n(s)\|_{1/2}^2 + c\left(\int_0^t \|w_n(s)\|_{3/2}^2\,\mathrm{d}s\right)^2$$

for some $c > 0$. Recombining these estimates of I_0 and multiplying by 2, (8.17) becomes

$$\sup_{s\in[0,t]}\|w_n(s)\|_{1/2}^2 + 2\int_0^t \|w_n(s)\|_{3/2}^2\,\mathrm{d}s$$

$$\leq a_1\int_0^t \|v_n(s)\|_1^4\,\mathrm{d}s + a_2\left(\int_0^t \|w_n(s)\|_{3/2}^2\,\mathrm{d}s\right)^2 \qquad (8.18)$$

$$+ a_3\int_0^t \left(\int_0^s \|u_n(r)\|_{1/2}^2\,\mathrm{d}r + \|u_0\|_{L^1}\right)^4\,\mathrm{d}s,$$

where $a_1, a_2, a_3 > 0$ are independent of n and t. To simplify the last term we fix $c' > 0$ such that

$$\int_0^t \left(\int_0^s \|u_n(r)\|_{1/2}^2 \, dr \right)^4 ds$$

$$\leq c't \left(\int_0^t \|v_n(s)\|_{1/2}^2 \, ds \right)^4 + c't^5 \sup_{s \in [0,t]} \|w_n\|_{1/2}^8.$$

Thus (8.18) becomes

$$\sup_{s \in [0,t]} \|w_n(s)\|_{1/2}^2 + 2 \int_0^t \|w_n(s)\|_{3/2}^2 \, ds$$

$$\leq a_1 \int_0^t \|v(s)\|_1^4 \, ds + a_2 \left(\int_0^t \|w_n(s)\|_{3/2}^2 \, ds \right)^2 + a_3 c't \|u_0\|_{L^1}^4 \quad (8.19)$$

$$+ a_3 c't \left(\int_0^t \|v(s)\|_{1/2}^2 \, ds \right)^4 + a_3 c't^5 \sup_{s \in [0,t]} \|w_n(s)\|_{1/2}^8.$$

This used the fact that $\|v_n(t)\|_\sigma$ is an increasing function of n for all $\sigma \geq 0$ and $t \in [0, T^n]$.

We next use (8.19) to find a uniform lower bound on the maximal existence time T_n, of u_n. It suffices to consider the case $T_n < \infty$. Comparing the w_n terms on the left-hand and right-hand sides of (8.19), we define

$$E(t) := a_2 \left(\int_0^t \|w_n(s)\|_{3/2}^2 \, ds \right) + a_3 c't^5 \sup_{s \in [0,t]} \|w_n(s)\|_{1/2}^6$$

and set

$$\tau_n := \sup \left\{ t \in [0, T_n) : E(t) \leq 1 \right\}.$$

Observe that $\tau_n < T_n$ since E is continuous and $E(t) \to \infty$ as $t \to T_n$, because $\|w_n(t)\|_{L^2}$ must blow up as $t \to T_n$. This also means that $E(\tau_n) = 1$.

As notation for the terms in the right-hand side of (8.19) that do not depend on w_n, we define

$$F(t) := a_1 \int_0^t \|v(s)\|_1^4 \, ds + a_3 c't \left(\int_0^t \|v(s)\|_{1/2}^2 \, ds \right)^4 + a_3 c't \|u_0\|_{L^1}^4.$$

Note that $F(t)$ is a continuous increasing function that is positive except at $t = 0$ (assuming that u_0 is non-zero). We now define

$$T := \sup \left\{ t \in [0, \infty) : F(t) < \min \left(\frac{1}{(16a_3c't^5)^{1/3}}, \frac{1}{2a_2} \right) \right\}.$$

It is easy to see that $T > 0$ and is independent of n. We will show that $T_n \geq T$ for all n. Suppose, for contradiction, that $\tau_n < T$, then by (8.19),

$$\frac{1}{2} \sup_{s \in [0, \tau_n]} \|w_n(s)\|_{1/2}^2 + \int_0^{\tau_n} \|w_n(s)\|_{3/2}^2 \, ds \leq F(\tau_n).$$

Hence

$$E(\tau_n) = a_2 \left(\int_0^{\tau_n} \|w_n(s)\|_{3/2}^2 \, ds \right) + a_3 c' \tau_n^5 \sup_{s \in [0, \tau_n]} \|w_n(s)\|_{1/2}^6 < 1.$$

This is a contradiction since we showed that $E(\tau_n) = 1$.

We have shown that $T_n \geq T$ for all n. Furthermore, arguing as above we have

$$\frac{1}{2} \sup_{s \in [0, T]} \|w_n(s)\|_{1/2}^2 + \int_0^T \|w_n(s)\|_{3/2}^2 \, ds \leq F(T).$$

Thus (u_n) is uniformly bounded in $L^2(0, T; H^{3/2})$ and $L^\infty(0, T; H^{1/2})$; moreover this regularity implies that $\partial_t u_n \in L^2(0, T; H^{-1/2})$, by a routine calculation. Proceeding as before with a standard compactness argument one can show that u is a local strong solution in the sense of (8.5).

Next we prove that this local solution is unique (this argument also applies to give the uniqueness we claimed in Section 8.2). Suppose that u and v are strong solutions to (8.5) with the same initial data. Set $w = u - v$ then taking the product of the equation satisfied by w with $2\Lambda^1 w$ yields the estimate

$$\|w(t)\|_{1/2}^2 + 2 \int_0^t \|w(s)\|_{1/2}^2 \, ds \leq c \int_0^t \|u(s)\|_{L^6} \|w(s)\|_1 \|w(s)\|_{3/2} \, ds$$

$$+ c \int_0^t \|w(s)\|_{H^{1/2}} \|v(s)\|_{3/2} \|w(s)\|_{3/2} \, ds. \tag{8.20}$$

For the first term we use interpolate $\|w\|_1^2 \leq \|w\|_{1/2} \|w\|_{3/2}$ and Young's inequality to obtain:

$$c\|u(s)\|_{L^6} \|w(s)\|_1 \|w(s)\|_{3/2} \leq c\|u(s)\|_{H^1}^4 \|w(s)\|_{1/2}^2 + \|w(s)\|_{3/2}^2. \tag{8.21}$$

For the second we make use of Lemma 8.2.1 and the fact that $w(0) = 0$:

$$c\|w(s)\|_{H^{1/2}}\|v(s)\|_{3/2}\|w(s)\|_{3/2} \le c\|v(s)\|_{3/2}^2\|w(s)\|_{1/2}^2 + \|w(s)\|_{3/2}^2$$

$$+ c\|v(s)\|_{3/2}^2 \left(\int_0^s \|w(r)\|_{1/2} \left(\|u(r)\|_{1/2} + \|v(r)\|_{1/2}\right) \,\mathrm{d}r\right)^2.$$

$$(8.22)$$

The integral over $[0, t]$ of the last term in (8.22) is at most

$$c\left(\int_0^t \|v(s)\|_{3/2}^2\mathrm{d}s\right)\left(\int_0^t \|w(s)\|_{1/2}^2\mathrm{d}s\right)\left(2\int_0^t \|u(s)\|_{1/2}^2 + \|v(s)\|_{1/2}^2\mathrm{d}s\right).$$

As $u \in L^4(0, T; H^{1/2})$ and $v \in L^2(0, T; H^{1/2}) \cap L^2(0, T; H^{3/2})$, this together with (8.20), (8.21) and (8.22) imply that

$$\|w(t)\|_{1/2}^2 \le \int_0^t G(s)\|w(s)\|_{1/2}^2 \,\mathrm{d}s$$

for some $G \in L^1(0, T)$. A Gronwall inequality now implies that, since $\|w(0)\|_{1/2} = 0$, $\|w(t)\|_{1/2} = 0$ for all $t \in [0, T]$. Uniqueness now follows using Lemma 8.2.1.

We have proved the following.

Lemma 8.3.1. *For $u_0 \in H^{1/2}$ there exists $T > 0$ and a unique strong solution $u \in L^2(0, T; H^{3/2}) \cap C^0([0, T]; H^{1/2})$ to the Burgers equations, in the sense of (8.5).*

Fix a representative of u that is continuous with respect to time into $H^{1/2}$. For almost every $t \in [0, T]$, we certainly have $u(t) \in H^1$, in which case we can apply Theorem 8.2.2 to obtain global classical solutions (on (t, ∞)) with initial data $u(t)$. By continuity of u and uniqueness of local strong solutions these classical solutions agree with u on their common domain. This completes the proof of Theorem 8.1.2.

8.4 Conclusions

We have shown that in the case of periodic boundary conditions the vector-valued diffusive Burgers equations have a unique solution given initial data in $H^{1/2}$. These solutions become classical immediately after the initial time and can be extended globally.

The results here contrast with classical results about the Navier–Stokes equations, which have thus far only been shown to have *local* well-posedness in $\dot{H}^{1/2}$ (see Marín-Rubio et al. (2013) or Chemin et al. (2006)). The main difference between these two systems seems to be the maximum principle for the Burgers equations. In other respects the analysis is slightly more straightforward in the case of Navier–Stokes, since we can make use of incompressibility.

In several places we appealed to the analysis of Fourier series but otherwise we have not used the periodicity of the solution in an essential way. Therefore we might expect similar results to hold on $\mathbb{R}^3$ or on other domains.

As discussed in the introduction we have not been able to find weak solutions for less regular data $u_0 \in L^2$ and it would be interesting to seek well-posedness of the Burgers equations in the various critical spaces that are often used to find local well-posedness results for the Navier–Stokes equations. Some examples of such spaces are: L^3 (Kato (1984)), certain Besov spaces (Cannone, Meyer, & Planchon (1994)) and BMO^{-1} (Koch & Tataru (2001)).

Irrespective of any approach in L^3 and the other aforementioned spaces, the existence of a maximum principle leads us to ask whether initial data $u_0 \in L^\infty \cap L^2$ is enough to deduce local or global well-posedness. The main obstacle to doing this seems to be that we must find classical solutions before applying the maximum principle, since the maximum principle does not seem to pass to the Galerkin approximations. Using another system of approximations might avoid this difficulty, for example, a variation on the time-discretisation approach of Kiselev & Ladyzhenskaya (1957).

References

Calderón, C.P. (1990) Existence of weak solutions for the Navier–Stokes equations with initial data in L^p. *Trans. Amer. Math. Soc* **318**, 179–200.

Cannone, M., Meyer, Y., & Planchon, F. (1994) Solutions auto-similaires des équations de Navier–Stokes. *Séminaire sur les Équations aux Dérivées Partielles,* 1993–1994, Exp. No. VIII, 12pp., École Polytech., Palaiseau, 1994.

Chemin, J.Y., Desjardins, B., Gallagher, I., & Grenier, E. (2006) *Mathematical Geophysics: an introduction to rotating fluids and the Navier–Stokes equations.* Oxford lecture series in mathematics and its applications, vol. 32. Oxford: Oxford University Press.

Constantin, P. & Foias, C. (1988) *Navier–Stokes Equations.* Chicago: The University of Chicago Press.

Evans, L.C. (2010) *Partial differential equations: second edition.* Providence R.I.: American Mathematical Society.

Fabes, E.B., Jones, B.F., & and Rivière, N.M. (1972) The initial value problem for the Navier–Stokes equations with data in L^p. *Arch. Rational Mech. Anal.* **45**, 222–240.

Galdi, G.P. (2000) An introduction to the Navier–Stokes initial-boundary value problem. *In:* Galdi, G.P., Heywood, J.G., & Rannacher, R. (eds), *Fundamental directions in mathematical fluid mechanics.* Birkhäuser-Verlag.

Kato, T. (1984) Strong L^p solutions of the Navier–Stokes equations in R^m with applications to weak solutions. *Mathematische Zeitschrift* **187**, 471–480.

Kiselev, A. & Ladyzhenskaya, O. (1957) On the existence and uniqueness of the solution of the nonstationary problem for a viscous, incompressible fluid. (Russian). *Izv. Akad. Nauk SSSR. Ser. Mat.* **21**, 655–680.

Koch, H. & Tataru, D. (2001) Well-posedness for the Navier–Stokes equations. *Advances in Mathematics* **157**, 22–35.

Ladyzhenskaya, O.A., Solonnikov, V.A., & and Ural'ceva, N.N. (1968) *Linear and quasilinear equations of parabolic type.* Translations of Mathematical Monographs, vol. 23. Providence, R.I.: American Mathematical Society. Translated from the Russian by S. Smith.

Marín-Rubio, P., Robinson, J.C., & Sadowski, W. (2013) Solutions of the 3D Navier–Stokes equations for initial data in $H^{1/2}$: Robustness of regularity and numerical verification of regularity for bounded sets of initial data in H^1. *J. Math. Anal. Appl.* **400**, 76–85.

Robinson, J.C. (2001) *Infinite-Dimensional Dynamical Systems.* Cambridge University Press.

Robinson, J.C. (2006) The 3d Navier–Stokes equations. *Bol. Soc. Esp. Mat. Apl. Sēma* **35**, 43–71.

Robinson, J.C., Rodrigo, J.L., & Sadowski, W. (2016) *The three–dimensional Navier–Stokes equations. Classical theory.* Cambridge University Press.

9

On the Fursikov approach to the moment problem for the three-dimensional Navier–Stokes equations

James C. Robinson

Mathematics Institute, University of Warwick,
Coventry, CV4 7AL. UK.
`j.c.robinson@warwick.ac.uk`

Alejandro Vidal-López

Department of Mathematical Sciences,
Xi'an Jiaotong-Liverpool University,
Suzhou 215123, China P. R.
`Alejandro.Vidal@xjtlu.edu.cn`

Abstract

In a series of papers Fursikov proposed a programme, based on analysing the moments of measure-valued statistical solutions, to obtain the density of initial data for which the corresponding solution of the 3D Navier–Stokes equations is regular. We illustrate the key points of the argument, and discuss some limitations of their method, by applying it to the case of the simple ODE $\dot{x} + x - x^2 = 0$ for which we know that such a density result cannot hold.

9.1 Introduction

Despite intensive efforts, the regularity problem for the three-dimensional Navier–Stokes equations is still unresolved, and is therefore very inter-

Published in *Recent Progress in the Theory of the Euler and Navier-Stokes Equations*, edited by James C. Robinson, José L. Rodrigo, Witold Sadowski, & Alejandro Vidal-López. ©Cambridge University Press 2016.

esting to consider alternative approaches to this well-established problem. In this paper we discuss one such approach due to Fursikov (1984, 1986a & b, 1987; see also Vishik & Fursikov, 1988), which treats not the equations themselves, but rather the linear 'Chain of Moments' (CoM) equations that arise when one considers the moments of statistical (measure-valued) solutions. If successful, the Fursikov programme would show that the set of initial data that give rise to regular solutions of the three-dimensional Navier–Stokes equations forms a dense subset of some sufficiently regular Sobolev space.

One can define a statistical solution by taking a measure on the space of initial data and then letting this evolve in a natural way under the flow induced by the equations. The moments of the resulting time-dependent measure satisfy a chain of infinitely-many linear equations, the 'Chain of Moments' (Section 9.3.1), and with enough regularity this implication can be reversed (Section 9.3.3). Proving the existence of 'regular' solutions to these equations is complicated by the fact that solving for the kth moment requires knowledge of the $(k+1)$th moment. Therefore Fursikov adopts a more roundabout approach, based on considering the equations within a variational framework and minimising a functional that penalises non-regular solutions: in this way they show that the CoM equations have a regular solution for a dense set of initial moments (Section 9.4). The essential difficulty in deducing a result on the 'density of regularity' lies in relating these approximate moments to a positive measure on the space of initial data, which we discuss further in Section 9.5.

In the context of the Navier–Stokes equations the arguments used to prove these results are extremely technical. The purpose of this paper is to discuss the many interesting ideas that arise as part of this theory, and to illustrate them more simply by applying them to the scalar ODE

$$\dot{x} + x - x^2 = 0$$

that shares some basic structural similarities with the Navier–Stokes equations: a linear dissipative term and a quadratic nonlinearity (which in the case of this ODE we know leads to the blowup of many solutions in a finite time).

9.2 A description of the Fursikov result

In this section we give a somewhat 'impressionistic' description of the result presented in Vishik & Fursikov (1988), preferring not to give the many technicalities required even in the construction of the appropriate spaces in which to treat the problem. We give many of the details in full for our simple example, which should help the interested reader who wishes to consider the original problem further.

We are concerned with the Navier–Stokes equations

$$\partial_t u - \Delta u + (u \cdot \nabla)u + \nabla p = 0, \qquad \nabla \cdot u = 0$$

with prescribed initial data $u(0) = u_0$ in a 'sufficiently smooth' Sobolev space H^α.

Given a probability measure μ_0 on H^s reflecting the distribution of the initial condition, we can construct a statistical solution of the Navier–Stokes equations. By this, we mean a probability measure on H^α defined by

$$\mu_t(\omega) := \mu_0(\{u_0 \in H : u(t; u_0) \in \omega\}), \quad \text{for all} \quad \omega \in \mathcal{B}(H^\alpha),$$

where $\mathcal{B}(H^\alpha)$ denotes the Borel sets of H^α. Such statistical solutions can be constructed, for instance, as limits of sequences of measures on finite-dimensional spaces using Galerkin-like approximations (see Vishik & Fursikov, 1988).

With an appropriate definition of the moments of a measure on H^α (the kth moment is an element of the k-fold tensorial product of H^α) one can then show that the time-dependent moments M_k of such a statistical solution satisfy the linear Chain of Moments (CoM) equations,

$$\dot{M}_k(t) + A_k M_k(t) + B_k M_{k+1}(t) = 0, \qquad M_k(0) = m_k^0, \qquad \text{(CoM)}$$

where m_k^0 are the moments of the initial probability distribution μ_0 and A_k and B_k are linear operators. The fact that this family of equations is linear is offset by the fact that we need to solve this system of equations 'from $k = \infty$', that is, in order to be able to find the kth moment we need to know the $(k+1)$th one.

If we consider the moments of a probability measure on H^α with support on the ball centred at 0 and radius R, the norms of these moments have at most a polynomial growth which depends on R; the space H_R^α will denote the space of moments with this growth. The space Y_R^α denotes the space of time-dependent moments that are 'regular'; regular solutions of (CoM) arising from the moments m^0 of a probability measure are the moments of the corresponding statistical solution and are supported on regular solutions of the Navier–Stokes equations (see Section 9.3.3 for more details in a simpler setting).

The main result in Vishik & Fursikov (1988) regarding the solvability of the chain of moments corresponding to the Navier–Stokes equations is that there is a dense set of initial moments for which (CoM) has a unique regular solution.

Theorem 9.2.1. *Let $R > 0$. For any initial sequence of moments $m^0 \in H_R^\alpha$, $\alpha > 2$, there exists a family of initial values $m^\epsilon \in H_R^\alpha$ such that*

(i) *for each $0 < \epsilon < 1$, the Chain of Moments (CoM) with initial value m^ϵ has a unique regular solution $M^\epsilon \in Y_R^\alpha$, and*

(ii) *$m^\epsilon \to m^0$ in H_ρ^α, with $\rho > R$, as $\epsilon \to 0$.*

Given an initial measure, the previous result allows us to construct a family of sequences of moments converging to those of the initial measure for which the corresponding solutions of (CoM) are unique (and regular). The problem that remains is to reconstruct a family of measures from this family of approximating moments. If we could do this, then by considering the support of these measures we could prove the density of initial data for which the Navier–Stokes equations have a regular solution. We discuss this in more detail in Section 9.5, but very briefly the idea is as follows: given an initial condition u_0, apply Theorem 9.2.1 with initial moments those of a delta distribution at u_0; the supports of the measures corresponding to the approximating moments should converge to that of δ_{u_0}, and hence there should be initial conditions arbitrarily close to u_0 for which there exits regular solutions. The possibility of deducing such a strong result via the Chain of Moments equations serves as a significant spur to their further study.

The positivity of the measures corresponding to the approximating moments is the key missing ingredient in the programme, and is essential in order to be able to localise the support of these measures, as well as to ensure the regularity of the statistical solution (which implies the regularity of the solutions on which it is supported); we discuss this further in Section 9.5.1, and then in Section 9.5.2 discuss some possible approaches that could circumvent this difficulty.

9.3 Statistical solutions and their moments

In order to illustrate the main ideas of the Fursikov program we use the simple ODE

$$\dot{x} + x = x^2, \qquad x \in \mathbb{R}, \tag{9.1}$$

158   *J.C. Robinson & A. Vidal-López*

together with the corresponding Cauchy problem

$$\dot{x} + x = x^2, \qquad x(0) = x_0 \in \mathbb{R}, \tag{9.2}$$

for $t \geq 0$. The Cauchy problem can be solved explicitly using separation of variables to yield

$$S(t)x_0 := x(t; x_0) = \frac{x_0}{x_0 + (1 - x_0)e^t}, \tag{9.3}$$

valid for all $t \geq 0$ if $x_0 \leq 1$, and for $0 \leq t < \log(x_0/(x_0 - 1))$ if $x_0 > 1$.

9.3.1 Statistical solutions and a formal derivation of the Chain of Moments equations

We begin by taking a 'statistical' approach: given an initial measure μ_0 on $\mathbb{R}$, the statistical solution of problem (9.1) is given by the time-dependent probability measures μ_t on $\mathbb{R}$ defined by setting

$$\mu_t(\omega) := \mu_0(S(t)^{-1}\omega) \tag{9.4}$$

for every $\omega \in \mathcal{B}(\mathbb{R})$, where $\mathcal{B}(\mathbb{R})$ denotes the Borel sets of $\mathbb{R}$ and $S(t)x_0$ denotes the solution of (9.1) at time t as in (9.3).

For a measure μ on $\mathcal{B}(\mathbb{R})$, its *moment of order k* (or *kth moment*) is given by

$$m_k = \int_{\mathbb{R}} x^k \, d\mu(x).$$

We then say that the sequence $m = m(\mu) = (m_k)_k$ is the *sequence of moments* corresponding to the measure μ.

We can now introduce the associated Chain of Moments (CoM) equations associated to (9.1). Given a measure μ_0, we denote its moments by m_k^0 and assume them to be finite for every $k \geq 0$; we want to write down the equations satisfied by the moments $M_k(t)$ of μ_t, the statistical solution of (9.1) defined in (9.4). [We use M_k to denote time-dependent moments, and m_k to denote the moments of a fixed measure, e.g. of an initial condition.]

Multiplying equation (9.1) by x^{k-1}, and integrating with respect to the measure μ_0 we obtain

$$\frac{1}{k}\frac{d}{dt}\int_{\mathbb{R}} x^k(t; x_0) \, d\mu_0 + \int_{\mathbb{R}} x^k(t; x_0) \, d\mu_0 - \int_{\mathbb{R}} x^{k+1}(t; x_0) \, d\mu_0 = 0,$$

where $d\mu_0$ is a measure on the initial condition x_0. After a change of variables we get

$$\frac{1}{k}\frac{\mathrm{d}}{\mathrm{d}t}\int_{\mathbb{R}} y^k \,\mathrm{d}\mu_t(y) + \int_{\mathbb{R}} y^k \,\mathrm{d}\mu_t(y) - \int_{\mathbb{R}} y^{k+1}\,\mathrm{d}\mu_t(y) = 0;$$

this yields the CoM equations

$$\dot{M}_k + kM_k - kM_{k+1} = 0, \qquad (9.5)$$

with initial data

$$M_k(0) = m_k^0. \qquad (9.6)$$

Note that these equations are linear, but solving for M_k requires knowledge of M_{k+1}. Before we try to solve the CoM equations, we first consider the space in which we will look for solutions.

9.3.2 The Fursikov spaces of moments

As we have already mentioned, the space of moments with polynomial growth plays an important role in the Fursikov argument. In the context of a measure on $\mathbb{R}$, this space is given by

$$H_R := \left\{ (m_k)_k,\ m_k \in \mathbb{R} : \ \|m\|_{H_R}^2 = \sum_{k\geq 1} R^{-2k}|m_k|^2 < \infty \right\}, \qquad (9.7)$$

endowed with the scalar product $(m, m')_{H_R} = \sum_{k\geq 1} R^{-2k} m_k m'_k$. In addition to this scalar product, we consider the duality action

$$\langle m', m \rangle = \sum_{k\geq 1} m'_k m_k$$

so that $H'_R = H_{1/R}$. Notice that weak convergence in H_R is equivalent to termwise convergence of the moments (i.e. $M^n \rightharpoonup M$ iff $M_k^n \to M_k$ for every k).

The choice of this space H_R is in part justified by the following lemma.

Lemma 9.3.1. *A positive finite measure μ with moments in H_R is supported on $[-R, R]$, $\mu(\{R\}) = \mu(\{-R\}) = 0$, and $|m_k(\mu)| < R^k \mu(\mathbb{R})$ for any $k \geq 1$. In particular, for a probability measure with moments in H_R, $|m_k(\mu)| < R^k$ for any $k \geq 1$.*

Proof. Suppose there exists a set of positive μ-measure $A \subset \mathbb{R} \setminus (-R, R)$. Then, for any k, the $2k$-moment of μ satisfies

$$m_{2k}(\mu) = \int_{\mathbb{R}\setminus(-R,R)} x^{2k}\,\mathrm{d}\mu \geq R^{2k}\mu(A) > 0,$$

with a similar lower bound if $\mu(\{-R,R\}) \neq 0$. Hence,

$$\|m(\mu)\|_{H_R}^2 = \sum_{k \geq 1} R^{-2k} |m_k(\mu)|^2 \geq \sum_{k \geq 1} R^{-4k} |m_{2k}(\mu)|^2 \geq \sum_{k \text{ even}} \mu(A) = \infty,$$

which contradicts the fact that $m(\mu) \in H_R$.

The inequality

$$|m_k| \leq \int_{[-R,R]} |x|^k \mathrm{d}\mu \leq R^k \mu(\mathbb{R}).$$

is now immediate.   □

The next result relates the convergence of moments to the convergence of measures, and shows that the convergence $m^\epsilon \rightharpoonup m^0$ implies, in particular, the convergence of the supports of μ^ϵ towards that of μ^0.

Proposition 9.3.2. *Let μ^ϵ, $\epsilon \geq 0$, be probability measures with moments $m^\epsilon \in H_R$. Then, $m^\epsilon \rightharpoonup m^0$ in H_R, i.e, $m_k^\epsilon \to m_k$ for $k \geq 1$, if and only if $\mu^\epsilon \rightharpoonup \mu^0$ as $\epsilon \to 0$.*

In this case, the support of the measure μ^ϵ converges to that of μ^0, in the sense that for any $x_0 \in \mathrm{supp}\,(\mu^0)$ there exist $x_\epsilon \in \mathrm{supp}\,(\mu^\epsilon)$ such that $x_\epsilon \to x_0$ as $\epsilon \to 0$.

Proof. First, if $\mu_\epsilon \rightharpoonup \mu^0$ then by definition

$$m_k^\epsilon = \int_{\mathbb{R}} x^k \,\mathrm{d}\mu^\epsilon(x) \to \int_{\mathbb{R}} x^k \,\mathrm{d}\mu^0(x) = m_k^0.$$

Suppose now that $m^\epsilon \rightharpoonup m^0$. Notice that from the assumptions we have that for every $\epsilon > 0$, $m^\epsilon \in H_R^+$ and so $\mathrm{supp}(\mu^\epsilon) \subset [-R,R]$. In particular, for any $k \geq 0$ we have

$$\int_{\mathbb{R}} x^k \,\mathrm{d}\mu^\epsilon(x) \to \int_{\mathbb{R}} x^k \,\mathrm{d}\mu^0(x) \quad \text{as} \quad \epsilon \to 0, \tag{9.8}$$

and so, for any polynomial $P(x)$ we have

$$\int_{\mathbb{R}} P(x) \,\mathrm{d}\mu^\epsilon(x) \to \int_{\mathbb{R}} P(x) \,\mathrm{d}\mu^0(x) \quad \text{as} \quad \epsilon \to 0.$$

Now, the measures μ^ϵ are supported on $[-R,R]$. So, the previous expression is equivalent to

$$\int_{[-R,R]} P(x) \,\mathrm{d}\mu^\epsilon(x) \to \int_{[-R,R]} P(x) \,\mathrm{d}\mu^0(x) \quad \text{as} \quad \epsilon \to 0$$

for any polynomial $P(x)$. Since the set of polynomials is dense in the space of continuous functions in $[-R, R]$, for any $f \in C([-R, R])$,

$$\int_{\mathbb{R}} f(x) \, d\mu^\epsilon(x) = \int_{[-R,R]} f(x) \, d\mu^\epsilon(x)$$

$$\to \int_{[-R,R]} f(x) \, d\mu^0(x) = \int_{\mathbb{R}} f(x) \, d\mu^0(x)$$

as $\epsilon \to 0$; and so $\mu^\epsilon \rightharpoonup \mu^0$ as $\epsilon \to 0$.

For the convergence of the support, if $x_0 \in \operatorname{supp}(\mu^0)$ then given any $\epsilon > 0$, we have $\mu^0(B_{\epsilon/2}(x_0)) > 0$ where $B_r(x_0)$ denotes the ball of radius r centred at x_0. Now take $f \in C^0(\mathbb{R})$ to have support in $B_\epsilon(x_0)$ and to be 1 on $B_{\epsilon/2}(x_0)$. Then from the weak convergence of the measures it follows that

$$\mu_\epsilon(B_\epsilon(x_0))) \geq \int_{\mathbb{R}} f \, d\mu^\epsilon \to \int_{\mathbb{R}} f \, d\mu^0 \geq \int_{B_{\epsilon/2}(x_0)} d\mu^0 > 0.$$

Thus, for ϵ small enough, there exists $x_\epsilon \in \operatorname{supp}(\mu^\epsilon) \cap B_\delta(x_0)$. Now, taking $\delta \to 0$ we have the desired convergence. $\qquad\square$

9.3.3 Regular solutions of the CoM give rise to regular statistical solutions

Given a measure μ^0 on the space of initial conditions, the formal manipulations in Section 9.3.1 suggest that the moments of the resulting statistical solution will be given by the solution of (9.5) with initial data $M_k(0) = m_k(\mu^0)$. In order to show that this is indeed the case we use a result guaranteeing the uniqueness of 'regular' solutions of the CoM equations; this can be viewed as an analogue of the well known uniqueness result for sufficiently regular solutions of the 3D Navier–Stokes equations.

We say that a time-dependent moment sequence $M = (M_k)$ is *regular* if it belongs to the space

$$Y_R(T) := \{M \in L^2_R(0, T; +) : \ \dot{M} \in L^2_R(0, T; -)\}, \tag{9.9}$$

where

$$L_R^2(0,T;\pm) := \{(M_k)_k,\ M_k \in L^2(0,T):$$

$$\|M\|_{L_R^2(0,T;\pm)}^2 = \sum_{k \geq 1} R^{-2k} k^{\pm 1} \|M_k\|_{L^2(0,T)}^2 < \infty\}.$$

The Sobolev-type embedding

$$Y_R(T) \subset C([0,T]; H_R),$$

where H_R was defined in (9.7), shows that this is a natural setting for the problem[1].

Our aim in this section is to prove the following result.

Theorem 9.3.3. *Let μ be a probability measure on $\mathcal{B}(\mathbb{R})$ with moments $m \in H_R$. Suppose that the CoM equations (9.5) with initial data m have a regular solution $M \in Y_R(T)$. Then, for any $t \in (0,T]$, the moments $\{M_k(t)\}_k$ correspond to those of the measure μ_t defined by*

$$\mu_t(\omega) := \mu(S(t)^{-1}\omega), \quad \omega \in B(\mathbb{R}), \tag{9.10}$$

i.e. they are the moments of the statistical solution with initial data μ. Furthermore this statistical solution is supported on $[-R,R]$, i.e. on bounded solutions of the ODE (9.1).

First we show that regular solutions of the CoM equations are unique if there are regular solutions of an appropriate adjoint problem, and then demonstrate the existence of such solutions. We then use this uniqueness to show that any regular solution of the Chain of Moments must in fact be the moments of the corresponding statistical solution. In the context of the Navier–Stokes equations the results of this section were proved by Fursikov (1987) (see also Vishik & Fursikov, 1988).

Theorem 9.3.4 (Following Theorem 3.1 in Fursikov, 1987). *If*

$$\begin{cases} \dot\Phi_1 - \Phi_1 = 0, & t \in [0,T_0] \\ \dot\Phi_k - k\Phi_k + (k-1)\Phi_{k-1} = 0, & t \in [0,T_0] \\ \Phi_k(T_0) = \Psi_k \end{cases} \tag{9.11}$$

[1]Notice also that (9.5) can be rewritten as

$$k^{-1/2}\dot M_k + k^{1/2}M_k - k^{1/2}M_{k+1} = 0$$

showing that $k^{-1/2}\dot M_k$ and $k^{1/2}M_k$ are quantities 'of the same order'. In the case of the Navier–Stokes equations, the $L^2(\pm)$ spaces represent those in which the nonlinear term can be well understood when considering the associated CoM equations: the $+/-$ represents a gain/loss of one derivative in one of the spaces used to construct the tensor product space where the moments live (see Vishik & Fursikov, 1988).

has a solution $\Phi \in Y_{1/R}(T_0)$ *for a dense set of* Ψ *in* $H_{1/R}$ *then any solution* $M \in Y_R(T)$ *of the CoM equations* (9.5) *with* $m \in H_R$ *is unique (among all solutions in* $Y_R(T)$*).*

Proof. Assume that $M \in Y_R(T)$ is a solution of (9.5) with initial data $m = 0$. We will show that $M(t) \equiv 0$ for all $t \in (0, T)$. Fix $\tau \in (0, T_0]$ and suppose that $\Phi \in Y_{1/R}(T_0)$ is a solution of (9.11). Define

$$\Phi_k^\tau(t) := \Phi_k(t + T_0 - \tau), \qquad 0 \le t \le \tau,$$

so that Φ^τ satisfies (9.11) on $[0, \tau]$ with $\Phi_k^\tau(\tau) = \Psi_k$.

Now multiply the equation for M_k,

$$\dot{M}_k + kM_k - kM_{k+1} = 0,$$

by Φ_k^τ, integrate from 0 to τ, and sum in k to obtain

$$0 = \sum_{k \ge 1} \int_0^\tau (\dot{M}_k + kM_k - kM_{k+1}) \Phi_k^\tau \, dt$$

$$= \sum_{k \ge 1} \int_0^\tau \dot{M}_k \Phi_k^\tau + kM_k \Phi_k^\tau - kM_{k+1} \Phi_k^\tau \, dt$$

$$= \sum_{k \ge 1} M_k(\tau) \Phi_k^\tau(\tau) - \sum_{k \ge 1} m_k \Phi_k^\tau(0)$$

$$- \int_0^\tau \left\{ \sum_{k \ge 1} (M_k \dot{\Phi}_k^\tau - kM_k \Phi_k^\tau) + \sum_{k \ge 2} (k-1) M_k \Phi_{k-1}^\tau \right\} \, dt.$$

Hence

$$0 = \sum_{k \ge 1} M_k(\tau) \Phi_k^\tau(\tau) - \sum_{k \ge 1} m_k \Phi_k^\tau(0)$$

$$- \sum_{k \ge 2} \int_0^\tau M_k(\dot{\Phi}_k^\tau - k\Phi_k^\tau + (k-1)\Phi_{k-1}^\tau) \, dt$$

$$- \int_0^\tau M_1(\dot{\Phi}_1^\tau - \Phi_1^\tau) \, dt.$$

Since $m = 0$ and Φ^τ satisfies (9.11) with $\Phi_k^\tau(\tau) = \Psi_k$, it follows that

$$\sum_{k \ge 1} M_k(\tau) \Psi_k = 0.$$

Since this holds for a dense set of Ψ, $M(\tau) = 0$; since $\tau \in (0, T_0]$ was arbitrary, $M(\tau) = 0$ for all $\tau \in [0, T_0]$. We can now repeat the argument in $[T_0, 2T_0]$, using the fact that $M(T_0) = 0$; in this way it follows inductively that $M(t) = 0$ for all $t \in [0, T]$, which yields the required uniqueness result. $\square$

The uniqueness of regular solutions of the CoM equations is now a consequence of the following existence result.

Theorem 9.3.5. *Given $0 < T_0 \leq \infty$, there exists a solution $\Phi \in Y_{1/R}(T_0)$ of the adjoint problem (9.11) for a dense set of initial moments m in $H_{1/R}$.*

To prove this result one can try to solve the adjoint problem directly, forward in k, and check the regularity of the solution. However, this approach only works when R is sufficiently small. Nevertheless, such a solution does in fact exist, as the following argument shows. We only sketch the proof, which is based on the fact that solutions of the underlying equation are analytic functions of the initial data.

Proof. (Sketch) Given an analytic function

$$\Psi(x) = \sum_{k=0}^{\infty} \Psi_k x^k, \quad x \in \mathbb{R},$$

the function

$$\Phi(x, t) = \Psi(S(T - t)x), \quad x \in \mathbb{R}, \quad 0 < t < T,$$

is a first integral of the equation (9.1), that is, $\Phi(x(t), t)$ is constant whenever $x(t)$ is a solution of (9.1).

Given an initial sequence of moments $m \in \bigcap_{R>0} H_R$, which is a dense subset of H_ρ for any choice of $\rho > 0$, we can define

$$\Psi(x) := \sum_{k=0}^{\infty} m_k x^k, \quad x \in \mathbb{R}.$$

This function will be analytic because of the growth restrictions on m_k imposed by the condition on m. We can then set

$$\Phi(x, t) := \Psi(S(T - t)x) = \sum_{k=0}^{\infty} m_k [S(T - t)x]^k,$$

which will be constant along trajectories and is another analytic function of x, which can be written as the power series

$$\Phi(x,t) = \sum_{k=0}^{\infty} \Phi_k(t)x^k.$$

Relatively straightforward calculations show that the coefficients Φ_k must satisfy the adjoint problem (9.11). It now only remains to check the regularity of the Φ_k; this can be done using the fact that the first integral and its derivatives with respect to both t and x are analytic on a ball centred at zero and radius R taking values in $C([0,T_0])$ for some $T_0 > 0$ small enough. $\qquad\square$

The existence of such a solution to the adjoint problem guarantees the uniqueness of regular solutions of the CoM equations. Given this, we can now prove (following Theorem 6.1 in Fursikov, 1987) the correspondence between statistical solutions and those of the CoM equations stated as Theorem 9.3.3.

Proof of Theorem 9.3.3. Notice that for any $R > 0$, there exists a time $T_0 > 0$ small enough such that the solution of the ODE (9.1) starting at $|u_0| \le R$, does not blow-up on $[0,T_0]$. In fact, we can choose T_0 so that the solutions starting at any point in $[-R,R]$ remain uniformly bounded on this time interval by some constant $\rho \ge R$.

Now, for any $t \in [0,T_0]$, the measure μ_t given in (9.10) is well defined, and has moments $M_k'(t)$ such that $M_k' \in Y_\rho$. Thus, M' satisfies the chain of moments (9.5) with the same initial condition as $M \in Y_R \subset Y_\rho$. Now, by Theorem 9.3.5, the solution of (9.5) is unique in Y_ρ and so, $M' = M$, i.e. the moments $M(t)$ in the statement of the theorem are the moments of μ_t for all $t \in [0,T_0]$. Since $Y_R \subset C([0,T_0]; H_R)$ we have $M(t) \in H_R$, for any $t \in [0,T_0]$. By Lemma 9.3.1, the measure $\mu(t)$ is concentrated in the ball B_R, and this proves the theorem on $[0,T_0]$.

Now, since $M(T_0) \in H_R$ and T_0 does not depend on u_0 (but only on R), the argument above can be carried out on $[T_0, 2T_0]$ with $\{M_k(T_0)\}$ as initial data and μ_{T_0} as initial measure, defining the measure μ for $t \in [T_0, 2T_0]$ as

$$\mu_{T_0+s}(\omega) = \mu_{T_0}(S(s)^{-1}\omega) = \mu_{T_0}(S(T_0)^{-1}S(s)^{-1}\omega) = \mu_{T_0}(S(T_0+s)^{-1}\omega).$$

We can repeat the argument until, after a finite number of steps, we find a measure μ_t defined for every $t \in [0,T]$. The statement about the support of this solution follows from Lemma 9.3.1. $\qquad\square$

Note that in the context of the Navier–Stokes equations Theorem 9.3.3 provides the main justification for treating the CoM equations. If we can prove the existence of a regular solution to these equations, this leads to a statistical solution supported on regular solutions of the Navier–Stokes equations themselves. In the next section we therefore turn to the problem of proving the existence of solutions of the CoM equations.

9.4 Regular solutions of the CoM equations for a dense set of initial moments

In this section we review the key ingredients of the arguments from Vishik & Fursikov (1988) and Fursikov (1987, 1986a, 1984) used to show that the CoM equations have regular solutions for a dense set of initial data in H_R. We apply these argument to the CoM equations

$$\dot{M}_k + kM_k - kM_{k+1} = 0, \qquad M_k(0) = m_k, \tag{9.12}$$

arising from the model ODE (9.1) and prove the following result.

Theorem 9.4.1. *Given any $m_0 \in H_R$, there exists a sequence $m^\epsilon \to m^0$ in H_ρ, for any $\rho > R$ as $\epsilon \to 0$ such that the solution M of the CoM equations (9.5) with $M(0) = m^\epsilon$ is regular, i.e. $M \in Y_R$.*

Proof. As we have already remarked, the idea is to treat the CoM equations (9.5) as an ill-posed problem: we look for regular solutions by solving a sequence of minimisation problems penalising non-regular solutions. In particular, we will consider the following family of problems indexed by $\epsilon > 0$,

$$(P)_\epsilon \begin{cases} \text{Find } (y, M) \in (H_R, L^2_R(0, T; +)) \text{ such that} \\ \frac{1}{2}\|y - m^0\|^2_{H_R} + \frac{\epsilon}{2}\|M(t)\|^2_{L^2_R(0,T;+)} \to \inf, \\ \dot{M}_k + kM_k - kM_{k+1} = 0, \quad \text{and} \\ M_k(0) = y_k, \end{cases}$$

where $m^0 \in H_R$ are the moments of a chosen initial measure μ_0 on $\mathbb{R}$. Note that we ask for time-dependent M_k that satisfy the CoM equations for initial data y which may not be equal to m^0; we try to minimise the difference between y and m^0 while requiring the solutions to be regular (note that if $M \in L^2_R(0, T; +)$ and $\dot{M}_k = -kM_k + kM_{k+1}$ then necessarily $\dot{M} \in L^2_R(0, T; -)$ and so $M \in Y_R(T)$, see (9.9)). If there is a regular solution M of (9.12) with initial moment m^0 then (m^0, M) is the minimiser.

Let us define $X_\epsilon := H_R \times L_R^2(0, T; +)$ with norm

$$\|(m, M)\|_\epsilon^2 = \|m\|_{H_R}^2 + \epsilon \|M\|_{L_R^2(0,T;+)}^2,$$

and $Z = H_R \times L_R^2(0, T; -)$ with norm

$$\|(m, M)\|_Z^2 = \|m\|_{H_R}^2 + \|M\|_{L_R^2(0,T;-)}^2.$$

The constraint in the minimisation problem is given by the kernel of the linear operator

$$U : H_R \times L_R^2(0, T; +) \to H_R \times L_R^2(0, T; -)$$

defined as

$$U(m, M) := (\{M_k(0) - m_k\}_k, \{\dot{M}_k + kM_k - kM_{k+1}\}_k).$$

The space of regular solutions of the chain of moments (9.5) is given by

$$Y_R(T) \subset C_R([0, T]; H_R) =: C_R(0, T).$$

Notice that $U : H_R \times Y_R \subset X_\epsilon \to H_R \times L_R^2(0, T; -)$ is continuous. Also, $U : X_\epsilon \to Z$ is an unbounded closed linear operator with domain

$$Y_{R,R} := \{(u_0, u) \in X_\epsilon : \dot{u} \in L_R^2(0, T; -), \ u(0) = u_0\},$$

which is dense in X_ϵ and is a subset of $H_R \times C_R(0, T)$. We endow $Y_{R,R}$ with the scalar product

$$(u, v)_{Y_{R,R}} = \sum_{k \geq 1} R^{-2k} u_k(0) v_k(0) + \sum_{k \geq 1} k R^{-2k} \int_0^T u_k(t) v_k(t) \, \mathrm{d}t$$

$$+ \sum_{k \geq 1} R^{-2k} k^{-1} \int_0^T \dot{u}_k(t) \dot{v}_k(t) \, \mathrm{d}t.$$

We can now rewrite the problem $(P)_\epsilon$ as

$$(P)_\epsilon \begin{cases} \text{Find } (y, M) \in X_\epsilon \text{ such that} \\ J(y, M) = \dfrac{1}{2} \|(y - m_0, M)\|_{X_\epsilon}^2 \to \inf \\ U(y, M) = 0. \end{cases}$$

Since the set of admissible elements (initial moments for which the corresponding solution of the CoM equations is regular) is not empty and the functional J is strictly convex, for any $\epsilon > 0$, $R > 0$, and $f \in [0, R]$ the problem $(P)_\epsilon$ has a solution $(y^\epsilon, M^\epsilon) \in Y_{R,R}$, which is unique in the class $H_R \times L_R^2(0, T; +)$.

In order to show that $y^\epsilon \to m^0$ as $\epsilon \to 0$ we want to use the Euler–Lagrange equations. However, we cannot guarantee that these hold for $(P)_\epsilon$ as it stands. While we know that the Lagrangian is given by

$$\mathcal{L}(x, v, f) = \frac{1}{2}\|x - x_0\|_{X_\epsilon}^2 + \lambda(U(x), v)_Z,$$

where $x_0 = (m^0, 0) \in X_\epsilon$ and $\lambda \in \mathbb{R}$, we cannot ensure that $\lambda \neq 0$, since the Lagrange principle requires the range of the Gâteaux derivative of the operator U evaluated at the minimiser to be the full space Z (see Fursikov (2000) for example).

Nevertheless, since U is closed and its domain is dense, we can use the following trick, which provides a modified version of the Euler–Lagrange equations (see Theorem 6.3, p. 513, in Vishik & Fursikov, 1988). Notice that this trick requires the constraint to be linear and does not allow inclusion-type constraints (see Section 9.5.2 for more details).

Lemma 9.4.2. *There exists a sequence* $\{x_\delta\}$, *where* $x_\delta \in X_\epsilon$ *with*

$$\|x_\delta - x_0\|_{X_\epsilon}^2 \leq \delta,$$

such that the problem $(P)_\epsilon$ *with* x_0 *replaced by* x_δ *has the same minimiser* $\widehat{x}_\epsilon$. *For this new problem, there exists an element* $y_\delta \in Y_R$ *such that*

$$(\mathcal{L}_x(\widehat{x}_\epsilon, y_\delta, x_\delta), z)_{X_\epsilon} = 0 \qquad \text{for every} \quad z \in X_\epsilon.$$

To apply Lemma 9.4.2 to our problem, for each $\epsilon > 0$ we choose $x_\epsilon = (m^\epsilon, W^\epsilon)$, where this comes from Theorem 9.4.2 with $\delta = \sqrt{\epsilon}$. In particular, $m^\epsilon \to m^0$ in H_R and $W^\epsilon \to 0$ in $L_R^2(0, T; +)$.

The modified Lagrangian (with $\lambda = 1$) for $(P)_\epsilon$ is given by

$$\mathcal{L}(m, M, \psi, \phi)$$

$$= \left(\frac{1}{2} \sum_{k \geq 1} R^{-2k}|m_k - m_k^\epsilon|^2 + \frac{\epsilon}{2} \sum_{k \geq 1} R^{-2k}k \int_0^T |M_k(t) - W_k^\epsilon|^2 \, \mathrm{d}t \right)$$

$$+ \sum_{k \geq 1} R^{-2k}(M_k(0) - m_k)\psi_k$$

$$+ \sum_{k \geq 1} R^{-2k}k^{-1} \int_0^T (\dot{M}_k(t) + kM_k(t) - kM_{k+1}(t))\phi_k(t) \, \mathrm{d}t.$$

Lemma 9.4.2 guarantees the existence of a pair $(\psi, \phi) \in H_R \times L_R^2(0, T; -)$

for which the Euler–Lagrange equations hold, i.e.

$$0 = \mathcal{L}_m(m, M, \psi, \phi)[(p, q)] = \sum_{k \geq 1} R^{-2k}(m_k - f_k^\epsilon)p_k - \sum_{k \geq 1} R^{-2k} p_k \psi_k$$

and

$$0 = \mathcal{L}_M(m, M, \psi, \phi)[(p, q)]$$

$$= \epsilon \sum_{k \geq 1} R^{-2k} k \int_0^T (M_k(t) - W_k^\epsilon(t)) q_k(t) \, \mathrm{d}t$$

$$+ \sum_{k \geq 1} R^{-2k} \int_0^T \left[(k^{-1} \dot{q}_k(t) + q_k(t)) \phi_k(t) - R^2 q_k(t) \phi_{k-1}(t) \right] \mathrm{d}t$$

$$+ \sum_{k \geq 1} R^{-2k} q_k(0) \psi_k$$

for any $(p, q) \in H_R \times L_R^2(0, T; -)$, where (m, M) is the solution of the problem $(P)_\epsilon$.

From the first equation, choosing p such that the only nonzero term is the kth one, we have

$$R^{-2k}(m_k - f_k^\epsilon - \psi_k)p_k = 0,$$

from which it follows that $\psi_k = m_k - f_k^\epsilon$.

From the second equation, integrating by parts and taking (p, q) such that the only nonzero terms are those in the kth position it follows that ψ and ϕ satisfy

$$\begin{cases} \dot{\phi}_1 - \phi_1 &= \epsilon(M_1(t) - W_1^\epsilon) \\ \dot{\phi}_k - k\phi_k + R^2 k \phi_{k-1} &= \epsilon k(M_k(t) - W_k^\epsilon), \quad k \geq 2 \\ R^{-2k} k^{-1} \phi_k(T) &= 0 \\ R^{-2k} k^{-1} \phi_k(0) &= R^{-2k} \psi_k, \end{cases} \tag{9.13}$$

which are the Euler–Lagrange equations for the minimiser of $(P)_\epsilon$. This problem can be solved 'backwards in time' (from $t = T$ to $t = 0$) but 'forwards in k' (cf. the adjoint problem (9.11) used in the uniqueness argument of Section 9.3.3) to obtain estimates on

$$\phi_k(0) = k\psi_k = k(m_k - f_k^\epsilon),$$

which can be used to show that

$$m^\epsilon \to f \quad \text{in} \quad H_\rho, \quad \text{for any} \quad \rho > R.$$

Finally, notice that the solution of the CoM equations starting at m^ϵ is regular in the sense that it belongs to $Y_R(T)$.   □

9.5 Moments and measures, revisited

The idea to obtain the density of initial data for which the solutions of (9.2) are regular is the following. We take an initial condition x_0 and consider a delta distribution at this point, δ_{x_0}. We can apply Theorem 9.4.1 to the sequence $m^0 = m(\delta_{x_0}) = x_0^k$ to obtain the existence of a sequence of moments $m^\epsilon \in H_R$, with $m^\epsilon \to m^0$ as $\epsilon \to 0$ such that the solution of the CoM equations starting at m^ϵ is regular.

Now we would like to use the result of Proposition 9.3.2 to deduce from $m^\epsilon \to m^0$ that the support of the corresponding measures μ^ϵ must converge to that of m^0, i.e. to x_0, and hence that there must be initial data arbitrarily close to x_0 for which (9.1) has a bounded solution. However, we know that this is not the case for (9.1) if $x_0 > 1$.

The resolution of this seemingly contradictory conclusion is that Proposition 9.3.2 requires that the moments m^ϵ are the moments of a positive measure, and there is no reason for this to be the case. Indeed, the last paragraph shows that the moments m^ϵ cannot be the moments of a positive measure, since this would lead to a contradiction.

This observation leads to another, which is very telling. Another, much more straightforward, way to obtain a sequence of moments with exactly the same properties as those stated in Theorem 9.4.1 is simply to truncate the initial sequence of moments from the kth term on, i.e. given $\{m_k^0\}_k$, to consider the approximating sequence

$$m^n = (m_1^0, m_2^0, m_3^0, \ldots, m_n^0, 0, 0, \ldots).$$

(Exactly the same observation is true for the Navier–Stokes result of Theorem 9.2.1.) It is clear that the solution of the CoM equations starting with such initial moments will have a regular solution, since $M_k^n(t) \equiv 0$ for all $k > n$. It is easy to see in the simple case of a positive measure on the real line that the moments cannot eventually be zero, since in this case taking k sufficiently large would imply that

$$0 = m_{2k} = \int_{\mathbb{R}} x^{2k}\, d\mu(x),$$

and if μ is a positive measure then it must be supported at $x = 0$, and consequently all the moments would be zero.

How then, can one characterise moment sequences that correspond to positive measures?

9.5.1 Positive-definite real moments

There is in fact a well-known characterisation of the real sequences corresponding to moments of positive measures (see Theorem X.4 in Reed & Simon, 1975).

Theorem 9.5.1 (Hamburger). *A sequence of real numbers (m_n) represents the moments of a positive measure on $\mathbb{R}$ if and only if for all integers $N > 0$ and all $\beta_0, \beta_1, \ldots, \beta_N \in \mathbb{C}$,*

$$\sum_{i,j=0}^{N} \overline{\beta}_i \beta_j m_{i+j} \geq 0.$$

While this places limits on the sequences that can arise as moments of a positive measure, any sequence of real numbers can arise as the moments of a signed measure on $\mathbb{R}$.

Proposition 9.5.2. *Given any sequence (a_n), with $a_n \in \mathbb{R}$, there exists a measure μ, which can be chosen to have a C^∞ density function, such that $m_n(\mu) = a_n$ for all $n \geq 0$.*

To construct the measure we can proceed as follows (Berg, Christensen, & Jensen, 1979). First, using a result by Borel (see Trèves, 2006) we know that given (a_n) there exists a function $\varphi \in C^\infty(\mathbb{R})$ with compact support such that

$$\frac{\mathrm{d}^n}{\mathrm{d}x^n} \varphi(0) = \mathrm{i}^n a_n.$$

If we let g be the Fourier transform of φ, so that

$$\varphi(y) = \int_{\mathbb{R}} \mathrm{e}^{\mathrm{i}xy} g(x) \, \mathrm{d}x,$$

then

$$a_n = \mathrm{i}^{-n} \frac{\mathrm{d}^n}{\mathrm{d}x^n} \varphi(0) = \int_{\mathbb{R}} x^n g(x) \, \mathrm{d}x, \quad n \geq 0$$

and so $a_n = m_n(\mathrm{Re}\,(g(x))\,\mathrm{d}x)$.

As a consequence, for any sequence (a_n) there exist two positive measure μ^+, μ^- such that, (a_n) are the moments of the measure $\mu = \mu^+ - \mu^-$.

Notice that, in particular, any sequence can be decomposed as the difference of two positive-definite sequences. (This proof generalises to sequences in $\mathbb{R}^k$ for any k.)

It follows that each of the approximating moment sequences m^ϵ from Theorem 9.4.1 corresponds to a signed measure on the space of initial data; but it is not clear how to use this to obtain any useful information about the set of initial data giving rise to regular (bounded) solutions.

9.5.2 Alternative approaches

While the result of Theorem 9.4.1 initially appears a strong one, the problem that the approximating initial moments do not correspond to positive measures is a very serious problem, as we have now indicated. Indeed, any method that uses no properties that are particular to the Navier–Stokes equations and do not hold for our model ODE cannot hope to yield a 'density of regularity' result, since it does not hold for our model ODE. Therefore one obvious approach, namely to try to work not in H_R but in H_R^+, the space of 'positive-definite' moments, is bound to fail: parts of the programme carry over to this case unchanged – for example, the existence of a minimiser for a sequence of approximating problems is straightforward (since the set of positive-definite moments is a convex closed set of the space of moments), but although this minimiser exists, it is not possible to write down the Euler–Lagrange equations since the set of positive-definite moments is not rich enough. We could try to weaken the topology so that this set has nonempty interior would allow us to obtain the Euler–Lagrange equations in a weaker sense (see Fursikov, 2000). However, the set of positive moments has empty interior in any reasonable topology which would allow us to use the Euler–Lagrange equations to obtain the density in Theorem 9.2.1.

In fact, Fursikov (1986b) recognised this problem, and made the observation that while our model ODE blows up in a finite time for any real $x_0 > 1$, it has bounded solutions for all time for every $x_0 \in \mathbb{C} \setminus (1, \infty)$, since the denominator in the solution

$$x(t) = \frac{x_0}{x_0 + (1 - x_0)e^t}$$

has non-zero imaginary part for every $t \neq 0$. So the set of initial conditions leading to bounded solutions is certainly dense in $\mathbb{C}$.

So instead of trying to construct a real measure, Fursikov (1986a) took the *real* moments m^ϵ and from these constructed an analytic functional on the set of solutions of the CoM equations, which he then extended to a complex measure on the complex version of Y_R. The trace at zero of the resulting measure is therefore obtained using only the real moments. This trace is a complex measure μ, which can be decomposed as

$$\mu = \mu_\Re^+ - \mu_\Re^- + \mathrm{i}(\mu_\Im^+ - \mu_\Im^-),$$

where all of the measures on the right-hand side are positive real measures that are supported on initial data leading to regular solutions of the Navier–Stokes equations (see Fursikov, 1986a). However, it is not entirely clear how to interpret this result; in particular there is an indeterminacy in locating the support of such a measure, as shown by the following result due to Aizenberg & Zalcman (1995, Theorem 1′).

Theorem 9.5.3. *Let K a compact set in the complex plane that does not contain the origin and separates 0 from ∞. Then, given any sequence $\{a_n\}_n \subset \mathbb{R}$ such that*

$$\limsup_{n \to \infty} \sqrt[n]{|a_n|} < \min_{z \in K} |z|$$

there exists a complex measure μ, supported on K, such that

$$a_k = \int_{\mathbb{C}} z^k \, \mathrm{d}\mu(z) \qquad \text{for all} \quad k \geq 0.$$

To specify fully a measure on $\mathbb{C}$ we need to consider a matrix of moments instead of a sequence, i.e.

$$m_{j,k} = \int_{\mathbb{C}} z^j \bar{z}^k \, \mathrm{d}z.$$

In this case, the real moments (i.e. those corresponding to $k = 0$) only form one column of the matrix. One possible approach to extract some significant consequences from the Fursikov moment theory would be to work in the fully-complex setting from the beginning, using the characterisation of positive-definite complex moments given by Theorem 2.1 in Atzmon (1975) and Theorem 4.1 in Horn (1977). However, it seems that even in this setting the problem remains that the 'trick' of Lemma 9.4.2 cannot be used since the set of positive-definite moments is not sufficiently rich (i.e. is not open in any natural topology).

9.6 Conclusion

The Fursikov result on the density of initial moments for which the CoM equations has a unique regular solution is a striking result, and seems to come tantalisingly close to a prove of the density of regularity for the Navier–Stokes equations themselves. In the end, following this programme to its conclusion appears to be derailed by the difficulty of confining the analysis to sequences of moments that correspond to the physically relevant case of positive measures.

Acknowledgments

The work of JCR was supported by an EPSRC Leadership Fellowship EP/G007470/1; AVL was supported by this grant, and also by a Marie–Curie IEF 52078 ULD3DNSE. We would both like to thank Andrei Fursikov for many interesting and stimulating discussions.

References

Aizenberg, L. & Zalcman, L. (1995) Instability phenomena for the moment problem. *Ann. Scuola Norm. Sup. Pisa Cl. Sci.* **22**, 95–107.

Atzmon, A. (1975) A moment problem for positive measure on the unit disc. *Pacific J. Math.* **59**, 317–325.

Berg, C., Christensen, J.P.R., & Jensen, C.U. (1979) A remark on the multi-dimensional moment problem. *Math. Ann.* **243**, 163–169.

Fursikov, A.V. (1984) Solvability of a chain of equations for space-time moments. *Mat. Sb. (N.S.)* **125**, 306–331.

Fursikov, A.V. (1986a) Analytic functionals and unique solvability of quasilinear dissipative systems under almost all initial conditions. *Trudy Moskov. Mat. Obshch.* **49**, 3–55.

Fursikov, A.V. (1986b) Space-time moments and statistical solutions concentrated on smooth solutoins of the three-dimensional Navier–Stokes system or a quasilinear parabolic system. *Soviet Math. Dokl.* **29**.

Fursikov, A.V. (1987) On the uniqueness of the solution of a chain of moment equations that correspond to a three-dimensional Navier–Stokes system. *Mat. Sb. (N.S.)* **134**, 472–495.

Fursikov, A.V (2000) *Optimal control of distributed systems. Theory and applications.* Translations of Mathematical Monographs **187**. American Mathematical Society.

Horn, R.A. (1977) On the moments of complex measures. *Math. Z.* **156**, 1–11.

Trèves, F. (2006) *Topological vector spaces, distributions and kernels.* Dover Publications, Inc., Mineola, NY, 2006. Unabridged republication of the 1967 original.

Reed, M. & Simon, B. (1975) *Fourier analysis, self-adjointness.* Methods of Modern Mathematical Physics **2**. Academic Press.

Vishik, M.I. & Fursikov, A.V. (1988) *Mathematical problems of statistical hydromechanics.* Kluwer Academic Publishers.

Widder, D.V. (1941) *The Laplace transform.* Princeton University Press, Princeton, N.J.

10

Some probabilistic topics in the Navier–Stokes equations

Marco Romito

Dipartimento di Matematica,
Università di Pisa,
Largo Bruno Pontecorvo 5,
56127 Pisa, Italia.
romito@dm.unipi.it

Abstract

We give a short overview of some topics concerning the ways randomness can be added to the three-dimensional Navier–Stokes equations.

10.1 Introduction

The present work is a short overview of some results concerning the interactions between the analysis of the three-dimensional Navier–Stokes equations and the theory of probability. Our special choice of topics does not give, by any means, a complete picture of the state of the art on the subject and several interesting matters and papers have been just outlined or even completely omitted. In the choice of topics there is definitely a bias over the work by the author of the present paper.

The main theme here is to consider the different ways randomness can be added to the Navier–Stokes equations. For some of these ways there is a reasonable physical justification, or a mathematical explanation. These reasons will be given when appropriate. Randomness can be added to the various data of the equations

$$\begin{cases} \dot{u} + (u \cdot \nabla)u + \nabla p = \nu \Delta u \; (+ \dot{\eta}), \\ \operatorname{div} u = 0, \end{cases} \tag{10.1}$$

namely, the initial condition, the external forcing and the parameters (here the viscosity). We will mainly consider the equations in dimension three and give some details for the two-dimensional case when the analysis in 3D is impractical.

Published in *Recent Progress in the Theory of the Euler and Navier-Stokes Equations*, edited by James C. Robinson, José L. Rodrigo, Witold Sadowski, & Alejandro Vidal-López. ©Cambridge University Press 2016.

In Section 10.2 we discuss some results obtained when randomness is added at the level of the initial condition, for instance, results of almost-sure global existence in super-critical spaces, the evolution of the distributions when the equations are started with a random initial condition (statistical solutions), and invariant measures of the flow.

In Section 10.3 we consider the equations forced by Gaussian white noise. The literature on this subject is huge; we focus on the existence of solutions that constitute a Markov process, discuss some topics on uniqueness and blow-up, and prove the existence of densities for finite-dimensional functionals of the solutions, as a probabilistic type of regularity.

Section 10.4 deals with invariant measures for the stochastically forced equations. In a way this should have been part of the previous section, but by importance it has deserved a section by its own. Here we discuss existence, uniqueness and convergence towards an invariant measure, as well as the existence of explicit invariant measures.

Finally, in Section 10.5 we consider questions where randomness is more hidden. We consider probabilistic representation formulas for the solutions of (10.1). In the last part of the section the focus is in the interaction between the equations and statistics.

10.1.1 Notation and setting

In the rest of the paper we mainly focus on the Navier–Stokes equations with periodic boundary conditions, either without any external force or driven by Gaussian white noise. Most of the results may be or have already been extended to other boundary conditions, external non-random forces, etc. We do not give further details and refer the reader to the references.

We consider equation (10.1) with periodic boundary conditions on the d dimensional torus $\mathbb{T}_d = [0, 2\pi]^d$ (most of the time $d = 3$, when necessary $d = 2$).

Let H be the standard space of square summable, divergence-free vector fields, defined as the closure of divergence-free, periodic, smooth vector fields, with inner product $\langle \cdot, \cdot \rangle_H$ and norm $\| \cdot \|_H$. Define likewise V as the closure of divergence-free, periodic, smooth vector fields with respect to the H^1 norm, with scalar product $\langle \cdot, \cdot \rangle_V$ and norm $\| \cdot \|_V$

Let Π_L be the Leray projector, $A = -\Pi_L \Delta$ the Stokes operator, and denote by $(\lambda_k)_{k \geq 1}$ and $(e_k)_{k \geq 1}$ the eigenvalues and the corresponding orthonormal basis of eigenvectors of A. Define the space $V_\alpha = D(A^{\frac{\alpha}{2}})$ for $\alpha \in \mathbb{R}$. In particular, $V_0 = H$ and $V_1 = V$.

Define the bi-linear operator $B : V \times V \to V'$ as $B(u, v) = \Pi_L (u \cdot \nabla v)$, for $u, v \in V$. We recall that $\langle u_1, B(u_2, u_3) \rangle = -\langle u_3, B(u_2, u_1) \rangle$. We refer, for instance, to Temam (1995) for a detailed account of the above definitions.

When appropriate, we will consider the random forcing $\dot{\eta}$ in (10.1) as $\dot{\eta} = \mathcal{S} \, dW$, where W is a cylindrical Wiener process on H (and hence $\dot{W}$ is space-time white noise), and $\mathcal{S}$ is a linear bounded operator on H. The role of $\mathcal{S}$ is to colour the noise in space, to provide some space regularity. The covariance of the driving noise is then $\mathcal{S}\mathcal{S}^\star$. The term $\mathcal{S} \, dW$ can always be represented as

$$\mathcal{S} \, dW = \sum_n \sigma_n d\beta_n g_n,$$

where $(g_n)_{n \in \mathbb{N}}$ is an orthonormal basis of H of eigenvectors of $\mathcal{S}\mathcal{S}^\star$, $(\beta_n)_{n \in \mathbb{N}}$ are independent standard Brownian motions, and $(\sigma_n)_{n \in \mathbb{N}}$ are suitable coefficients.

10.2 Randomness in the initial condition

A natural way to include uncertainty in an evolution is to consider a probability distribution that weights the possible initial conditions. Moreover, the analysis of the evolution of distributions can give some knowledge of the dynamical properties of the system. For instance, a change of regularity of the measure might be interpreted in terms of the existence of different basins of attractions. Tao (2016) indicates the belief that blow-up for the three-dimensional Navier–Stokes equations might be more likely than regularity, but that carefully chosen initial distributions might avoid blow-up initial states and give only solution with global regularity. This should be an index of instability of blow-up with respect to small perturbations of the initial conditions (see also Section 10.3.5).

There has been recently a renewed interest in studying evolution equations with random initial condition, see for instance Burq & Tzvetkov (2008a,b, 2014). These ideas date back already to Bourgain (1996, 1994), who considers the space-periodic nonlinear Schrödinger equations in the focusing/defocusing case.

A common theme of these works is that typically, the random initial condition may provide a short time effect of smoothing by averaging that may overcome some obstructions due to the scaling invariance of the equations. This is the case when one can prove an almost sure (with respect to the probabilistic structure given by the initial distribution) existence of a local solution when starting from a super-critical space.

10.2.1 The randomization

Let us define the statistical distribution that has been used in the works we will be interested in. Let $\mathcal{H}$ be a Hilbert space and let $(e_n)_{n\geq 1}$ be an orthonormal basis of $\mathcal{H}$. Consider a sequence $(\Lambda_n)_{n\geq 1}$ of centred independent random variables with the property that there exists $c_1 > 0$ such that

$$\mathbb{E}[\Lambda_n^2] \leq c_1, \qquad \text{for every } n \geq 1.$$

Additional uniform moments (e.g. exponential) may provide additional properties or strong estimates. In this direction, a reasonable assumption due to Nahmod, Pavlović, & Staffilani (2013) is

$$\mathbb{E}[e^{\gamma\Lambda_n}] \leq e^{c_2\gamma^2}, \qquad \text{for every } \gamma \in \mathbb{R}, n \geq 1. \tag{10.2}$$

This, for instance, provides exponential concentration around the mean of the randomization we are going to define. Fix $f \in \mathcal{H}$, the 'seed', and define the *randomization* of f as

$$\Lambda f = \sum_n \Lambda_n f_n e_n,$$

where $(f_n)_{n\geq 1}$ are the Fourier coefficients of f with respect to the basis $(e_n)_{n\geq 1}$. It is immediate to see that Λf is a centred $\mathcal{H}$–valued random variable with covariance $\mathcal{U}_f^\star\mathcal{U}_f$, where $\mathcal{U}_f$ is the operator

$$x = \sum_n x_n e_n \qquad \leadsto \qquad \mathcal{U}_f x = \sum_n f_n x_n e_n.$$

If we choose, for instance, the initial random coefficients $(\Lambda_n)_{n\geq 1}$ as standard Gaussian, then Λf is also Gaussian with zero expectation and covariance $\mathcal{U}_f^\star\mathcal{U}_f$, and this characterise its distribution.

We first notice that the randomization does not give any gain in terms of smoothness. Mimicking Sobolev spaces, let us consider some compact subspace $\mathcal{H}_0$ of $\mathcal{H}$ defined by $\|x\|_{\mathcal{H}_0} = \sum \lambda_n^2 x_n^2 < \infty$, with $\lambda_n \uparrow \infty$.

Let $f \in H$ with $\|f\|_{\mathcal{H}_0} = \infty$. Let us prove that the randomization Λf cannot be in $\mathcal{H}_0$ almost surely. The proof is immediate in the Gaussian case, namely when the random variables $(\Lambda_n)_{n\geq 1}$ are standard Gaussians. Indeed,

$$\mathbb{E}[\|\Lambda f\|_{\mathcal{H}_0}^2] = \mathbb{E}\Big[\sum_n \lambda_n^2 \Lambda_n^2 f_n^2\Big] = \sum_n \lambda_n^2 f_n^2 = \infty,$$

and Fernique's theorem (see Bogachev, 1998) immediately implies that $\|\Lambda f\|_{\mathcal{H}_0} = \infty$, almost surely. In general, if the $(\Lambda_n)_{n\geq 1}$ are independent and uniformly not-too-often zero, then the same conclusion holds (Burq & Tzvetkov, 2008a, Lemma B.1). Since a way to use the randomization is to deduce improved summability of the solution of the linear problem (see for instance Proposition 10.2.3), another way to look at the lack of regularization is to recall that Besov spaces, and in turn Sobolev spaces, can be characterised in terms of regularity of the caloric extension (see for instance Lemarié-Rieusset, 2002).

We want to study now the support of the law of Λf. We recall that for a topological space E, endowed with the Borel σ-algebra, the (topological) support of a measure μ is the set of all points x such that $\mu(A) > 0$ for each neighbourhood of x.

Lemma 10.2.1. *Given $f \in \mathcal{H}$, the support of the law of Λf is the whole of $\mathcal{H}$ if and only if the support of the law of each Λ_n is $\mathbb{R}$ and $f_n \neq 0$ for every $n \geq 1$, where $(f_n)_{n\geq 1}$ are the Fourier coefficients of f with respect to the basis $(e_n)_{n\geq 1}$.*

Proof. Given $x \in \mathcal{H}$ and $\epsilon > 0$, we prove that $\mathbb{P}[\Lambda f \in B_\epsilon^{\mathcal{H}}(x)] > 0$. Let $N \geq 1$ and denote by $\Pi_{\leq N}$ and $\Pi_{>N}$ the projections, respectively, onto low and high modes. Choose N so that

$$\|\Pi_{>N} x\|_{\mathcal{H}} \leq \frac{\epsilon}{4}, \qquad \|\Pi_{>N} f\|_{\mathcal{H}} \leq \frac{\epsilon}{8},$$

then

$$\|\Lambda f - x\|_{\mathcal{H}} \leq \|\Pi_{\leq N}(\Lambda f - x)\|_{\mathcal{H}} + \|\Pi_{>N}\Lambda f\|_{\mathcal{H}} + \|\Pi_{>N} x\|_{\mathcal{H}}$$

$$\leq \|\Pi_{\leq N}(\Lambda f - x)\|_{\mathcal{H}} + \|\Pi_{>N}\Lambda f\|_{\mathcal{H}} + \frac{\epsilon}{4}.$$

Therefore, by the above estimate and using independence,

$$\mathbb{P}[\Lambda f \in B_\epsilon^{\mathcal{H}}(x)] = \mathbb{P}[\|\Lambda f - x\|_{\mathcal{H}} \leq \epsilon]$$

$$\geq \mathbb{P}\Big[\|\Pi_{\leq N}(\Lambda f - x)\|_{\mathcal{H}} \leq \frac{\epsilon}{2}, \|\Pi_{>N}\Lambda f\|_{\mathcal{H}} \leq \frac{\epsilon}{4}\Big]$$

$$= \mathbb{P}\Big[\|\Pi_{\leq N}(\Lambda f - x)\|_{\mathcal{H}} \leq \frac{\epsilon}{2}\Big] \mathbb{P}\Big[\|\Pi_{>N}\Lambda f\|_{\mathcal{H}} \leq \frac{\epsilon}{4}\Big] > 0,$$

since, on the one hand by the Chebychev inequality,

$$\mathbb{P}\left[\|\Pi_{>N}\Lambda f\|_{\mathcal{H}} \le \frac{\epsilon}{4}\right] = 1 - \mathbb{P}\left[\|\Pi_{>N}\Lambda f\|_{\mathcal{H}} \ge \frac{\epsilon}{4}\right] \ge$$

$$\ge 1 - \frac{16}{\epsilon^2}\boldsymbol{E}[\|\Pi_{>N}\Lambda f\|_{\mathcal{H}}^2] = 1 - \frac{16}{\epsilon^2}\|\Pi_{>N}f\|_{\mathcal{H}}^2 \ge \frac{3}{4},$$

on the other hand, by the independence and the assumption on the support of the $(\Lambda_n)_{n\ge 1}$,

$$\mathbb{P}\left[\|\Pi_{\le N}(\Lambda f - x)\|_{\mathcal{H}} \le \frac{\epsilon}{2}\right] \ge \mathbb{P}\left[|\Lambda_n f_n - x_n| \le \frac{\epsilon f_n}{2\|f\|_{\mathcal{H}}}, n \le N\right]$$

$$= \prod_{n\le N} \mathbb{P}\left[|\Lambda_n f_n - x_n| \le \frac{\epsilon f_n}{2\|f\|_{\mathcal{H}}}\right] > 0.$$

To prove the converse, notice that if $f_n = 0$ for some n, then the choice $x = e_n$ yields $\|\Lambda f - x\|_{\mathcal{H}} \ge |\Lambda_n f_n - 1| = 1$. Likewise, if the support of the law of Λ_1 is not $\mathbb{R}$, then there are $x_0 \in \mathbb{R}$ and $\epsilon_0 > 0$ such that $\mathbb{P}[\Lambda_1 \in (x_0 - \epsilon_0, x_0 + \epsilon_0)] = 0$. The choice $x = x_0 f_1 e_1$ yields $\|\Lambda f - x\|_{\mathcal{H}} \ge |f_1| \cdot |\Lambda_1 - x_0| \ge \epsilon_0 |f_1|$ almost surely. $\qquad\square$

Clearly the same proof of the lemma above holds true if $f \in \mathcal{H}_0$, for some subspace $\mathcal{H}_0$ of $\mathcal{H}$, namely in the latter case the support of the law of Λf is $\mathcal{H}_0$.

10.2.2 Strong local solution with random initial condition

We summarize how to show the existence of a local smooth solution with random initial condition in the energy space H, following Zhang & Fang (2011).

Let $u_0 \in H$ be the 'seed', and consider the random initial condition Λu_0, using the Fourier basis of H. The main idea is that there is an immediate gain of summability from L^2 to L^3. As L^3 is a critical space for Navier–Stokes, using results by Kato (1984) we know that there is a unique local solution.

Assume $\boldsymbol{E}[\Lambda_k^4] \le c$ for every $k \in \mathbb{Z}_\star^3$. Clearly, if stronger moments (e.g. exponential) are finite, the probability estimates below are better.

As mentioned above, through randomization of $u_0 \in H$ and the fourth moment condition for the randomizing variables, it follows that

$$\mathbb{E}[\|\Lambda u_0\|_{L^3}^3] \le c_3\|u_0\|_{L^2}^3.$$

Therefore the local existence and uniqueness of Kato (1984) yields the following result.

Theorem 10.2.2. *Let $u_0 \in H$, then with probability one there exist $T_\star = T_\star(\omega) > 0$ and a unique solution u with initial condition Λu_0, such that for all $p \geq 3$,*

- $t^{\frac{1}{2} - \frac{3}{2p}} u \in C([0, T_\star); L^p(\mathbb{T}_3))$,
- $t^{1 - \frac{3}{2p}} \nabla u \in C([0, T_\star); L^p(\mathbb{T}_3))$.

The time $T_\star = \infty$ on an event Ω_∞ with $\mathbb{P}[\Omega_\infty] \geq 1 - c_4 \|u_0\|_{L^2}^3$ (exponentially close to 1 with exponential moments of Λ_k).

Under the assumption of a finite sixth moment, a similar statement holds in $H^{\frac{1}{2}}$ (as in Fujita & Kato, 1964), namely with probability one there exists a unique solution u such that

$$u - e^{-tA}\Lambda u_0 \in C([0, T_\star]; V_{\frac{1}{2}}) \cap L^2(0, T_\star; V_{\frac{3}{2}}),$$

as well as an estimate of the probability that $T_\star \geq T$ (with $T \in (0, 1]$) in terms of $\|u_0\|_{L^2}$ (polynomial or exponential depending on the moments of the random coefficients of the randomization). Similar conclusions are given by Deng & Cui (2011a), proving that the solution is global if $\|u_0\|_{L^2}$ is small enough.

10.2.3 Global weak solutions with random initial conditions

The problem of finding an (interesting - see below in Section 10.2.5) initial distribution so that almost surely with respect to this distribution there is a unique global solution is still essentially open (but see Section 10.2.4 below). Clearly, there may be some 'trivial' example, such as some measure concentrated on small initial conditions in, for instance, $H^{1/2}$, but this adds nothing to what we know. A well supported initial distribution that gives rise, almost surely, to global strong solutions, would suggests that a blow-up in the equation is exceptional, or 'unstable', in the sense that a small variation in the initial condition might not lead to a singularity (more on this will be discussed in Section 10.3.5.3).

A way to obtain global *weak* solutions with no smallness assumption on the data, when starting from super-critical initial conditions has been recently proposed by Nahmod et al. (2013).

Here the 'smoothing' effect of the randomization they use is again in terms of summability of the solution of the linear problem. The exponential tail estimate is a consequence of assumption (10.2).

Proposition 10.2.3. *Let $\alpha \geq 0$, $p \geq 2$, $\sigma \geq 0$, and $\gamma \in \mathbb{R}$ satisfying $(\sigma + \alpha - 2\gamma)p < 2$, and $T > 0$, then for every $u_0 \in V_{-\alpha}$ we have*

$$\mathbb{P}[\|S_{\gamma,\sigma}u_0\|_{L^q(0,T;L^p(\mathbb{T}_3))} \geq \lambda] \leq c_5 e^{-c_6 \frac{\lambda^2}{c_7\|u_0\|_{-\alpha}^2}},$$

where $S_{\gamma,\sigma}u_0(t) = t^\gamma A^{\frac{\sigma}{2}} e^{-tA} \Lambda u_0$.

Let us define weak solutions in the following way.

Definition 10.2.4 (Definition 2.4, Nahmod et al., 2013). Given $\alpha > 0$, $u_0 \in V_{-\alpha}$, and a weak solution of the Navier–Stokes equations on $[0, T]$ is a vector field u such that,

- $u \in L^\infty_{\text{loc}}((0,T);H) \cap L^2_{\text{loc}}((0,T);V) \cap C((0,T);V_{-\alpha}^{\text{weak}})$,
- $u' \in L^1(0,T;V')$,
- the equation is satisfied in V', and
- $u(t) \to u_0$ weakly in $V_{-\alpha}$, as $t \to 0$.

The main theorem is as follows.

Theorem 10.2.5 (Thm. 2.6, Nahmod et al., 2013). *If $T > 0$, $\alpha \in (0, \frac{1}{4})$ and $u_0 \in V_{-\alpha}$, then with probability one there exists a global weak solution of (10.1) with initial condition Λu_0, of the form $u = e^{-tA}\Lambda u_0 + w$, where $w \in L^\infty_{\text{loc}}(0,\infty;H) \cap L^2_{\text{loc}}(0,\infty;V)$.*

In short, the idea behind the theorem is that one can use the smoothing effect of the randomization of the initial condition to produce a mild solution, defined for a short time. The solution immediately belongs to H and a standard weak solution can be started after a small time interval. It remains then only to show that the mild solution and the weak solution can be joined to obtain a weak solution as defined above.

This is the crucial point that forces the restriction $\alpha < \frac{1}{4}$ in the main theorem. Indeed, to prove the equivalence between weak and mild solutions (see Nahmod et al., 2013, Lemma 4.2, but similar assumptions are used in other crucial results of the paper) for the equation for $w = u - e^{-tA}\Lambda u_0$, where terms like $B(e^{-tA}\Lambda u_0, w)$ appear, one needs that, for instance $A^{\frac{1}{4}}e^{-tA}\Lambda u_0 \in L^{\frac{8}{3}}((0,T) \times \mathbb{T}_3)$. This happens, according to their lemma, when $\alpha < \frac{1}{4}$.

In the case $\alpha \in [\frac{1}{4}, 1)$ something can still be said, at least in terms of mild solutions.

Theorem 10.2.6 (Deng & Cui, 2011b). *Let $\alpha \in [\frac{1}{4}, 1)$ and $u_0 \in V_{-\alpha}$,* then with probability one there are $T = T(\omega)$ and a unique solution u of (10.1) on $[0, T]$ with initial condition Λu_0, such that

$$u - e^{-tA}\Lambda u_0 \in L^{\frac{4}{1+\alpha}}(0, T; L^{\frac{6}{1-\alpha}}).$$

Again, estimates of the probability that $T > t$ are available for small t, namely if $t \in (0, 1]$, there exists a unique solution on $[0, t]$ on an event Ω_t with

$$\mathbb{P}[\Omega_t] \geq 1 - c_8 e^{-\frac{c_9}{t^{\frac{1}{2}(1-\alpha)}\|u_0\|^2_{-\alpha}}}.$$

10.2.4 Fursikov's almost sure global well–posedness

The previous section detailed results where a suitable choice of the distribution of the initial condition would allow to prove existence of a strong (or weak) solution with supercritical data. It is well known that there exist initial distributions that ensure almost sure global well–posedness. This is a general result due to Fursikov (see Višik & Fursikov (1988) and also Fursikov, 1981b, 1983, 1984). Clearly, it is not difficult to provide initial distributions that give almost-sure global well-posedness; think of the Dirac at the origin, or some measure concentrated in a small ball of $V_{\frac{1}{2}}$. Indeed, the main problem of Fursikov's initial measures is that they are only characterized by their moments. It is well known that moments do not identify uniquely a measure (unless some growth condition holds). Moreover, no information on the support of these measure is available (unlike in the previous result; see Lemma 10.2.1). This is the reason for Fursikov to analyse the infinite dimensional system generated by the moments; see Fusikov (1987) and Višik & Fursikov (1988).

10.2.4.1 Statistical solutions

Statistical solutions were first introduced by Foias (1972) as a family, indexed by time, of probability measures satisfying the equations, appropriately recast (see also Foias et al., 2001). A different notion of statistical solution, seen as a measure on the path space (from this point of view, closer to the style of this paper; see Section 10.2.5 below), was formulated by Višik & Fursikov (1988). Let us consider the latter definition.

Define, for a fixed $s \gg 0$,

$$\mathcal{S}_T = \{u : u \in L^2(0, T; V) \cap L^\infty(0, T; H), \dot{u} \in L^\infty(0, T; V_{-s})\}$$

A *space-time statistical solution* with initial condition μ, is a probability measure $\mathbb{P}$ on $\mathcal{S}_T$ such that

- $\mathbb{P}[\mathcal{S}_T] = 1$,
- the marginal of $\mathbb{P}$ at time $t = 0$ is μ,
- $\mathbb{P}$ is concentrated on solutions of (10.1),
- for every $t \in [0, T]$,

$$\mathbb{E}^{\mathbb{P}}[\|u\|_{L^2(0,T;V)}^2 + \|u(t)\|_H^2 + \|u\|_{L^\infty(0,T;H)}^2 + \|\dot{u}\|_{L^\infty(0,T;V_{-s})}^2]$$

$$\leq c_{10}(1 + \mathbb{E}^{\mu}[\|x\|_H^2]).$$

A statistical solution represents the overall distribution of the stochastic process generated by the solutions of (10.1) when the initial distribution is given by the initial measure μ.

A variant of the definition above, more suited for the next section is as follows. Define

$$\mathcal{H}_T^{1,2} = \{u : u \in L^2(0, T : V_2), \dot{u} \in L^2(0, T : H)\}.$$

Notice that if $u \in \mathcal{H}_T^{1,2}$, then $u \in C([0, T]; V)$. A space-time statistical solution is a probability measure $\mathbb{P}$ on $\mathcal{H}_T^{1,2}$ such that for every $z \in L^2(0, T; H)$ and $\phi \in C_b(\mathcal{H}_T^{1,2})$,

$$\mathbb{E}^{\mathbb{P}}[\langle \dot{u} + \nu Au + B(u), z\rangle_{L^2(0,T;H)}\phi(u)] = 0.$$

10.2.4.2 Statistical extremal problems and a.s. smoothness

Given a measure μ on H such that $\boldsymbol{E}^{\mu}[\|x\|_H^k] < \infty$ for every $k \geq 1$, define its kth-moment M_k as the element of $\bigotimes_k H$ (the tensor product of k copies of H), such that

$$\langle M_k, \phi\rangle_{H,k} = \int \langle \otimes_k u, \phi\rangle_{H,k}\mu(du), \qquad \phi \in \bigotimes_k H,$$

where $\langle \cdot, \cdot \rangle_{H,k}$ and $\|\cdot\|_{H,k}$ are the tensorizations of the scalar product and the norm of H. The existence of M is granted by the Riesz representation theorem.

Consider a measure μ on H such that $\boldsymbol{E}[e^{\|x\|_H^2}] < \infty$ (so that all moments of μ are finite) and denote by $(m_k)_{k \geq 1}$ its moments). Consider the following functional defined over probability measures on $\mathcal{H}_T^{1,2}$,

$$\mathcal{J}(\mathbb{P}) = \mathbb{E}^{\mathbb{P}}[e^{\|u\|_{L^2(0,T;V_2)}^2}] + \rho\sum_{k=1}^{\infty}\frac{1}{k!}\|m_k - M_k\|_{H,k},$$

where $\rho > 0$ and $(M_k)_{k \geq 1}$ are the moments of the marginal of $\mathbb{P}$ at time

$t = 0$. For a measure $\mathbb{P}$ with $\mathcal{J}(\mathbb{P}) < \infty$, its moments do not grow too much (by the first term), and are not too different (at least at time 0) from the moments of μ.

The idea is to look at $\inf \mathcal{J}(\mathbb{P})$ over all statistical solutions $\mathbb{P}$ on $\mathcal{H}_T^{1,2}$. It turns out that the direct methods of the calculus of variations are effective. The functional is convex, lower semi-continuous in an appropriate topology and finite in at least one measure, so that the infimum is attained and there is a unique probability measure that realizes the minimum.

If we look at the marginal ν at time 0 of the unique minimizer, then for ν-almost every initial condition (10.1) has a unique global smooth solution.

Remark 10.2.7. Starting from (10.1), the system of equations

$$\dot{M}_k + \nu A_k M_k + B_k M_{k+1} = 0,$$

for the moments of a statistical solution can be derived, where A_k and B_k are suitable tensorizations of the Stokes operator A and the Navier–Stokes nonlinearity B.

Fursikov (see Fusikov, 1987, Višik & Fursikov, 1988 and related references) proves that there is a dense set of initial conditions for the moment system such that each of these initial conditions yields a unique solution. The problem here is that one cannot produce in general a statistical solution from moments.

Remark 10.2.8. In fact the first results of Fursikov (1980, 1981a) in this direction deal with the set of right-hand sides yielding a global, smooth solution. More precisely, set

$$N(u) = \dot{u} + \nu Au + B(u),$$

and solve $N(u) = f$, with a given initial condition $u(0) = u_0 \in V$. It turns out that the map $N : \mathcal{H}_T^{1,2} \to L^2([0,T];H)$ is continuous, so that the set $F_{u_0} = N(\{u \in \mathcal{H}_T^{1,2} : u(0) = u_0\})$ is exactly the set of all right-hand sides $f \in L^2(0,T;H)$ such that the Navier–Stokes equations with forcing f admit a unique smooth solution.

Moreover, F_{u_0} is open in the topology of $L^2(0,T;H)$ and dense in $L^2(0,T;H)$ with in the topology $L^p(0,T;V_{-\ell})$, for suitable p and ℓ. The result can be made independent from the viscosity and holds for the Euler equations, as long as the initial condition is smooth enough.

10.2.5 Invariant measures

The randomization of the initial condition to obtain local existence (in a super-critical space) or global existence (of a regular solution) becomes extremely useful when one knows that the system admits a (formal) invariant measure. Some explicit information is also required (the support of the measure, for instance).

So far, we do not know any explicit[1] invariant measure for the Navier–Stokes equations, and in fact we will have better luck with the randomly forced equation in Section 10.4.3.

We point out that a method to find invariant measures has been proposed by Foias et al. (2001) using generalized limits of time averages Here a generalized limit is any linear operator that extends the ordinary notion of limit. The existence of extensions is ensured by the Hahn–Banach theorem.

Euler is a different story, and indeed explicit invariant measures can be derived. In dimension three the known conserved quantities are: the kinetic energy, the helicity, namely $\int u \cdot \xi$, where $\xi = \operatorname{curl} u$ is the vorticity, the circulation, that is, $\int_{X_t(\gamma)} u(t) \cdot dx$, where γ is a curve in physical space and X is the flow induced by u, as well as the total momentum $\int u \, dx$ and angular momentum $\int x \cdot u \, dx$. The only good candidate then is the energy, and one can consider the Gibbs-like measures

$$\mu_{E,\beta} = \frac{1}{Z_\beta} e^{-\beta E(u)} \, du,$$

where $E(u) = \int |u|^2 \, dx$ is the energy. The above measure is interpreted as usual as a Gaussian measure. The problem is that such measures are supported on fairly large spaces, as

$$\int \|x\|^2_{-\alpha} \, \mu_\beta(dx) = \infty$$

for $\alpha \le \frac{3}{2}$. The problem of the existence of a flow of solutions of Euler which leaves the above measures invariant is (yet another) open problem.

The situation is much better in dimension two, due to weaker regularity requirements, but above all, due to the existence of a wealth of invariants, first of all the enstrophy $S(u) = \frac{1}{2} \int \xi^2 \, dx$, as well as $\int g(\xi) \, dx$ for every reasonable g. If J is any of the above invariants, the measure $Z_{J,\beta}^{-1} e^{-\beta J(u)} \, du$ would provide a formal invariant measure. The only reasonable measure though, those we may hope to give a sense, are given

---

[1]And interesting!, as otherwise any steady solution $\bar{u}$, including $\bar{u} = 0$ would provide the invariant measure $\delta_{\bar{u}}$.

by $J = E, S$. These measures are *infinitesimally invariant* in the sense that for every smooth function F depending only on a finite number of Fourier modes, $\int B(F) \, d\mu_{J,\beta} = 0$.

The measures originating from enstrophy have a smaller support, and hence it is expected that it should be easier to work with them. Indeed Albeverio & Cruzeiro (1990) showed that there exists a flow in $V_{-\alpha}$ (with $\alpha > \frac{1}{2}$) of Euler with invariant measure $\mu_{S,\beta}$. Measures from energy are more complicated. It even happens that $\int E(u) \, \mu_{S,\beta}(du) = \infty$, but the *renormalized energy* : $E := E - \mathbb{E}^{S,\beta}[E]$ (carefully interpreted as a limit of spectral approximations) makes sense, $e^{-\gamma:E:}$ is integrable with respect to $\mu_{S,\beta}$ and the measure $Z^{-1} e^{-\beta S - \gamma:E:}$ is again invariant. We refer the reader to Albeverio & Ferrario (2008) for more details. Here we raise the (philosophical) open problem of understanding the role of these invariant measures in connection with the physical phenomenon of turbulence.

We will see later that when adding a noise we will be able to find invariant measures for Navier–Stokes (in Sections 10.4.1 and 10.4.3). Ideas of *renormalization* will also play a significant role later, see Sections 10.4.3.1 and 10.4.3.2.

10.3 Randomness in the driving force

In the same way one can derive, at least formally, the Euler equations from the Lagrangian motion of fluid particles, a version of the Navier–Stokes equations driven by a special multiplicative noise depending on the gradient of the velocity can be derived starting from the Lagrangian motion perturbed by noise; see Brzeźniak, Capiński, & Flandoli (1991), Mikulevicius & Rozovski (2004) and Mikulevicius & Rozovskii (2005). The presence of random forcing can also take into account all those small fluctuations that affect the motion of a fluid and that are difficult to incorporate in a robust theory. We refer the reader to Flandoli et al. (2008) for the connections between the equations with random forcing and the theory of turbulence.

There is already a well-developed theory for stochastic PDEs, and in particular for equations from fluid dynamics. We refer to Flandoli & Gątarek (1995), Flandoli (2008) and Debussche (2013). Here we detail a recent approach initiated by Da Prato & Debussche (2003a) that looks for solutions with additional structure, the Markov property. For a well-posed problem the Markov property would not be an issue; on the other hand for problems where well-posedness is an open problem (as is in this work) extra care is needed.

There are at least two approaches that grant existence of Markov processes solving (10.1) driven by Gaussian noise. The first (see Da Prato & Debussche, 2003a, Debussche & Odasso, 2006) is essentially based on strong solutions of (10.1). In short the idea is to solve the Kolmogorov equation associated with the spectral Galerkin approximations. In order to grant the existence of a limiting object of the solutions of the Kolmogorov eqution, the authors look at a Kolmogorov equation perturbed by a strong potential. The solutions of the two equations are related by a Feynamn–Kac formula. The potential, a negative exponential of the H^2 norm, does an 'importance sampling' of strong solutions, since non-smooth solution would contribute with an infinite potential and hence with a null contribution in the Feynman–Kac formula.

The second approach (see Flandoli & Romito, 2006, 2008) is based on weak solutions. The construction builds over an abstract selection principle originally due to Krylov (1973) (see also Stroock & Varadhan, 1979). The idea is essentially to identify special classes of solutions, understood as probability measures on the space of trajectories, that are closed by conditional probability and for which weak-strong uniqueness holds. We refer also to Blömker, Flandoli, & Romito (2009) for another model where this theory can be applied (see also Blömker & Romito, 2009, 2012, 2013).

In this section we will consider (10.1) driven by a Gaussian noise, namely the noise $\dot{\eta} = \mathcal{S}\dot{W}$ in (10.1) is coloured in space by a covariance operator $\mathcal{S}^\star \mathcal{S} \in \mathscr{L}(H)$, where W is a cylindrical Wiener process (see Da Prato & Zabczyk (1992) for further details). We assume that $\mathcal{S}^\star \mathcal{S}$ is trace-class and we denote by $\sigma^2 = \mathrm{Tr}(\mathcal{S}^\star \mathcal{S})$ its trace. Finally, consider the sequence $(\sigma_k^2)_{k\geq 1}$ of eigenvalues of $\mathcal{S}^\star \mathcal{S}$, and let $(q_k)_{k\geq 1}$ be the orthonormal basis in H of eigenvectors of $\mathcal{S}^\star \mathcal{S}$. For simplicity we may assume that the Stokes operator A and the covariance commute, so that

$$\dot{\eta}(t,y) = \mathcal{S}\,\mathrm{d}W = \sum_{\mathbf{k}\in\mathbb{Z}^3_*} \sigma_{\mathbf{k}}\dot{\beta}_{\mathbf{k}}(t)e_{\mathbf{k}}(y).$$

10.3.1 Weak and strong solutions

Let us write (10.1), as usual, as an abstract stochastic equation,

$$\mathrm{d}u + (\nu Au + B(u))\,\mathrm{d}t = \mathcal{S}\,\mathrm{d}W, \qquad (10.3)$$

with initial condition $u(0) = x \in H$. A weak martingale solution is a filtered probability space $(\widetilde{\Omega}, \widetilde{\mathscr{F}}, \widetilde{\mathbb{P}}, \{\widetilde{\mathscr{F}}_t\}_{t \geq 0})$, a cylindrical Wiener process $\widetilde{W}$ on H and a process u with trajectories in $C([0, \infty); D(A)') \cap L^\infty_{\mathrm{loc}}([0, \infty), H) \cap L^2_{\mathrm{loc}}([0, \infty); V)$ adapted to $(\widetilde{\mathscr{F}}_t)_{t \geq 0}$ such that the above equation is satisfied with $\widetilde{W}$ replacing W.

Equivalently, a weak martingale solution can be described as a measure on the path space. Let $\Omega_{\mathrm{NS}} = C([0, \infty); D(A)')$ and let $\mathscr{F}^{\mathrm{NS}}$ be its Borel σ-algebra. Denote by $\mathscr{F}^{\mathrm{NS}}_t$ the σ-algebra generated by the restrictions of elements of Ω_{NS} to the interval $[0, t]$ (roughly speaking, this is the same as the Borel σ-algebra of $C([0, t]; D(A)')$). Let ξ be the canonical process, defined by $\xi_t(\omega) = \omega(t)$, for $\omega \in \Omega_{\mathrm{NS}}$.

Definition 10.3.1 (Flandoli & Romito, 2008). A probability measure $\mathbb{P}$ on Ω_{NS} is a solution of the martingale problem associated to (10.3), with initial distribution μ, if

- $\mathbb{P}[L^\infty_{\mathrm{loc}}(\mathbb{R}^+, H) \cap L^2_{\mathrm{loc}}(\mathbb{R}^+; V)] = 1$,
- for each $\phi \in D(A)$, the process

$$\langle \xi_t - \xi_0, \phi \rangle + \int_0^t \langle \xi_s, A\phi \rangle - \langle B(\xi_s, \phi), \xi_s \rangle \, \mathrm{d}s$$

is a continuous square summable martingale with quadratic variation $t\|\mathcal{S}\phi\|^2_H$ (hence a Brownian motion), and
- the marginal of $\mathbb{P}$ at time 0 is μ.

The second condition in the definition above has a two-fold meaning. On the one hand it states that the canonical process is a weak (in terms of PDEs) solution, and on the other hand it identifies the driving Wiener process, and hence is a weak (in terms of stochastic analysis) solution.

10.3.1.1 Strong solutions

It is also well-known that (10.3) admits local smooth solutions defined up to a random time (a stopping time, in fact) τ_∞ that corresponds to the (possible) time of blow-up in higher norms. To consider a quantitative version of the local smooth solutions, notice that τ_∞ can be approximated monotonically by a sequence of stopping times

$$\tau_R = \inf\{t > 0 : \|Au_R(t)\|_H \geq R\},$$

where u_R is a solution of the following problem,

$$\mathrm{d}u_R + \left(\nu Au_R + \chi(\|Au_R\|^2_H/R^2)B(u_R, u_R)\right) \mathrm{d}t = \mathcal{S}\,\mathrm{d}W,$$

with initial condition in $D(A)$, and where $\chi : [0, \infty) \to [0, 1]$ is a suitable *cut-off* function, namely a non-increasing, smooth function such that $\chi \equiv 1$ on $[0, 1]$ and $\chi_R \equiv 0$ on $[2, \infty)$. The process u_R is a strong

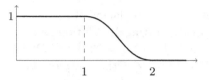

(in the PDE sense) solution of the cut-off equation. Moreover it is a strong solution also in terms of stochastic analysis, so it can be realized uniquely on any probability space, given the noise perturbation.

As it is well-known in the theory of Navier–Stokes equations, the regular solution is unique in the class of weak solutions that satisfy some form of the energy inequality. We will give two examples of such classes for Navier–Stokes with noise.

Remark 10.3.2. The analysis of strong (PDE meaning) solutions can be done on larger spaces, up to $D(A^{1/4})$, which is a critical space with respect to the Navier–Stokes scaling. The extension is a bit technical though; see Romito (2011a).

10.3.1.2 Solutions satisfying the almost sure energy inequality

An almost sure version of the energy inequality has been introduced by Romito (2008, 2010a). Given a weak solution $\mathbb{P}$, choose $\phi = e_k$ as a test function in the second property of Definition 10.3.1, to get a one dimensional standard Brownian motion β^k. Since $(e_k)_{k\geq 1}$ is an orthonormal basis, the $(\beta^k)_{k\geq 1}$ are a sequence of independent standard Brownian motions. Then the process $W_{\mathbb{P}} = \sum_k \beta^k e_k$ is a cylindrical Wiener process[1] on H. Let $z_{\mathbb{P}}$ be the solution to the linearization at 0 of (10.3), namely the solution of

$$\mathrm{d}z + Az = \mathcal{S}\,\mathrm{d}W, \qquad\qquad (10.4)$$

with initial condition $z(0) = 0$, and where $W = W_{\mathbb{P}}$. Finally, we set $v_{\mathbb{P}} = \xi - z_{\mathbb{P}}$. It turns out that $v_{\mathbb{P}}$ is a solution of

$$\dot{v} + \nu Av + B(v + z_{\mathbb{P}}, v + z_{\mathbb{P}}) = 0, \qquad \mathbb{P}\text{-a.s.},$$

---

[1]Notice that W is measurable with respect to the solution process.

with initial condition $v(0) = \xi_0$. An energy balance functional can be associated to $v_{\mathbb{P}}$,

$$\mathcal{E}_t(v,z) = \frac{1}{2}\|v(t)\|_H^2 + \nu \int_0^t \|v(r)\|_V^2 \, \mathrm{d}r - \int_0^t \langle z(r), B(v(r)+z(r), v(r)) \rangle_H \, \mathrm{d}r.$$

We say that a solution $\mathbb{P}$ of the martingale problem associated to (10.3) (as in Definition 10.3.1) satisfies the *almost sure energy inequality* if there is a set $T_P \subset (0, \infty)$ of null Lebesgue measure such that for all $s \notin T_P$ and all $t \geq s$,

$$P[\mathcal{E}_t(v,z) \leq \mathcal{E}_s(v,z)] = 1.$$

It is not difficult to check that $\mathcal{E}$ is measurable and finite almost surely.

10.3.1.3 A martingale version of the energy inequality

An alternative formulation of the energy inequality that, on the one hand is compatible with conditional probabilities, and on the other hand does not involve additional quantities (such as the processes $z_{\mathbb{P}}$ and $v_{\mathbb{P}}$) can be given in terms of super-martingales. The additional advantage is that this definition is amenable to generalization to state-dependent noise.

Define, for every $n \geq 1$, the process

$$\mathscr{E}_t^1 = \|\xi_t\|_H^2 + 2\nu \int_0^t \|\xi_s\|_V^2 \, \mathrm{d}s - 2\operatorname{Tr}(\mathcal{S}^\star \mathcal{S}),$$

and, more generally, for every $n \geq 1$,

$$\mathscr{E}_t^n = \|\xi_t\|_H^{2n}$$

$$+ 2n\nu \int_0^t \|\xi_s\|_H^{2n-2}\|\xi_s\|_V^2 \, \mathrm{d}s - n(2n-1)\operatorname{Tr}(\mathcal{S}^\star \mathcal{S}) \int_0^t \|\xi_s\|_H^{2n-2} \, \mathrm{d}s,$$

when $\xi \in L_{\mathrm{loc}}^\infty([0,\infty); H) \cap L_{\mathrm{loc}}^2([0,\infty); V)$, and ∞ elsewhere.

We say that a solution $\mathbb{P}$ of the martingale problem associated to (10.3) (as in Definition 10.3.1) satisfies the *super-martingale energy inequality* if for each $n \geq 1$, the process $\mathscr{E}_t^n$ defined above is $\mathbb{P}$-integrable and for almost every $s \geq 0$ (including $s = 0$) and all $t \geq s$,

$$\mathbb{E}[\mathscr{E}_t^n | \mathscr{F}_s^{\mathrm{NS}}] \leq \mathscr{E}_s^n,$$

or, in other words, if each $\mathscr{E}^n$ is an almost-sure supermartingale.

10.3.2 The selection principle

In order to carry on the construction of a Markov solution, we need to start with a class of solutions satisfying some minimal properties (sort of a set-valued Markov property). Given $x \in H$, let $\mathscr{C}(x) \subset \mathrm{Pr}(\Omega)$ be a set of weak martingale solutions (no other requirement so far) starting at x. The three main properties we shall require are:

- *Disintegration*: the classes $(\mathscr{C}(x))_{x \in H}$ are closed by conditional probabilities: if $\mathbb{P} \in \mathscr{C}(x)$, then the conditional probability distribution of $\mathbb{P}$ given $\mathscr{F}_t^{\mathrm{NS}}$ is in $\mathscr{C}(\omega(t))$ for $\mathbb{P}$ a.e. ω.
- *Reconstruction*: this is, in a way, the inverse operation of disintegration: if one has a $\mathscr{F}_t^{\mathrm{NS}}$ measurable map $x \mapsto \mathbb{Q}_x$, with $\mathbb{Q}_x \in \mathscr{C}(x)$, and $\mathbb{P} \in \mathscr{C}(x_0)$, then the probability measure given by $\mathbb{P}$ on $[0,t]$, and, conditionally on $\omega(t)$, by the values of $\mathbb{Q}.$, is an element of $\mathscr{C}(x_0)$.
- *Weak-strong uniqueness*: each solution coincides with the process u_R on $[0,t]$ on the event $\{\tau_R \geq t\}$.

For the construction of a Markov solution, we require the first two properties, the third one is necessary for further analysis (continuity with respect to the initial condition (see Section 10.3.3 below) and convergence to a unique invariant state; see Section 10.4.1).

The idea is to shrink each set $\mathscr{C}(x)$ to a single element by a series of reductions, while keeping the above properties. Fix a family $(\lambda_n, f_n)_{n \geq 1}$ which is dense in $[0, \infty) \times C_b(D(A)')$ and consider the functionals $J_n = J_{\lambda_n, f_n}$, where $J_{\lambda, f}$ is given by

$$J_{\lambda, f}(P) = \mathbb{E}^P \left[\int_0^\infty e^{-\lambda t} f(\xi_t) \, dt \right].$$

for arbitrary $\lambda > 0$ and $f : D(A)' \to \mathbb{R}$ upper semi-continuous. Next, set

$$\mathscr{C}_0(x) = \mathscr{C}(x), \qquad \mathscr{C}_n(x) = \{\mathbb{P} \in \mathscr{C}_{n-1}(x) : J_n(\mathbb{P}) = \sup_{\mathbb{Q} \in \mathscr{C}_{n-1}(x)} J_n(\mathbb{Q})\}.$$

All these sets are compact and their intersection is a single element (the selection associated to this maximised sequence), and therefore $\bigcap_{n \in \mathbb{N}} \mathscr{C}_n(x) = \{\mathbb{P}_x\}$.

Example 10.3.3. The existence of Markov solutions holds even without noise, when the solution is suitably understood as a probability on the space of trajectories. Consider the classical non-uniqueness example $\dot{X} = \mathrm{sgn}(X)\sqrt{|X|}$, with initial condition in $\mathbb{R}$. The problem has a unique solution $X_x(\cdot)$ for each initial condition $x \neq 0$, and two families of solutions $\{X_a^\pm = X_\star^\pm((t-a) \vee 0) : a \geq 0\}$ for the initial condition $x = 0$, where $X_\star^-, X_\star^+$ are the minimal and the maximal unique solution starting at 0.

If $\mathscr{C}(x)$ denotes the set of all solutions starting at x, viewed as probability measures on the path space $C([0,\infty); \mathbb{R})$, then $\mathscr{C}(x) = \{\delta_{X_x}\}$ for $x \neq 0$, where δ_{X_x} is the Dirac measure concentrated on X_x.

If $x = 0$, the solution starts at 0 and stays for an arbitrary time, then follows one of the solutions $X_\star^\pm$ (suitably translated). So the departing time from 0 can be interpreted as a random time T whose law can be arbitrary. Therefore any selection of solutions is completely described by a random time T and on $[0,\infty]$ and a coin flip C to decide to go up or down.

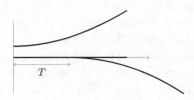

It is easy to see that a selection of solutions is Markov if and only if T is exponential (the lack of memory plays a major role), including the degenerate cases of infinite or zero rate (namely, $T = 0$ or $T = \infty$), with T and C independent.

Denote by $(\mathbb{P}_x^a)_{x \in [0,1]}$ the Markov families with rate a. We shall call *extremal* all those Markov solutions that can be obtained by the selection procedure. It turns out that the only extremal families are those corresponding to $a = 0$ and $a = \infty$.

In view of Section 10.3.3, we notice that no solution can be continuous with respect to the initial condition.

10.3.3 Continuity with respect to the initial condition

As we shall see, Markov solutions have a good structure, good enough to ensure that solutions are continuous (in an appropriate sense) with respect to the initial condition. In a way, for well-posedness we are only missing uniqueness.

Continuity with respect to the initial condition here is understood in terms of continuity of the law, in the total variation distance, of the solution for fixed time and seen as a function of the starting point. This is

M. Romito

a purely probabilistic notion and in fact it is ruled out for the equations
without noise, as it can be easily seen by the elementary consideration
shown in Figures 10.1 and 10.2. Without noise the 'law' of the solu-

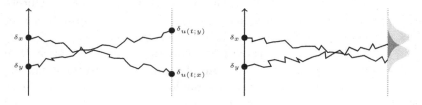

Fig. 10.1. Without noise...   Fig. 10.2. With noise...

tion evolves as a Dirac mass centred at the value of the solution and
no possible shrinking of the total variation distance is possible (unless
statistical solutions are considered, as in Section 10.2.4.1, but there is no
smoothing effect by the noise). With noise the two distributions have a
common mass for two reasons. The first reason is general: the diffusive
effect of the Gaussian perturbation; the second reason is due to weak-
strong uniqueness: there is a tiny (but non–zero) probability that τ_∞
may be large enough, so that the two laws are close.

We shall use the two facts above to show continuity. Indeed, for small
times, using weak-strong uniqueness,

$$\mathbb{E}[\phi(u(\epsilon;y))] - \mathbb{E}[\phi(u(\epsilon;x))] =$$
$$= \underbrace{\boldsymbol{E}[\phi(u(\epsilon;y))\mathbb{1}_{\{\tau_\infty>\epsilon\}}] - \mathbb{E}[\phi(u(\epsilon;x))\mathbb{1}_{\{\tau_\infty>\epsilon\}}]}_{\text{estimate with the regular solution}} +$$

$$+ \underbrace{\boldsymbol{E}[\phi(u(\epsilon;y))\mathbb{1}_{\{\tau_\infty\le\epsilon\}}] - \mathbb{E}[\phi(u(\epsilon;x))\mathbb{1}_{\{\tau_\infty\le\epsilon\}}]}_{\text{short time tail of }\tau_\infty}.$$

For short times the nonlinearity has a small effect, so that the dynamics
is essentially linear and the probability $\mathbb{P}[\tau_\infty \le \epsilon] \approx \mathrm{e}^{-1/\epsilon}$. On the other

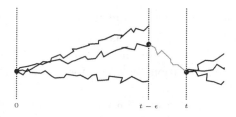

hand, when we are below τ_∞ we can work with the strong solutions (see Section 10.3.1.1) and hence with the classical theory. For times of order one, the real picture is that the 'uniqueness of strong solutions' argument is applied at the very last moment only, thanks to the Markov property,

$$P_t\phi(y) - P_t\phi(x) = P_\epsilon(P_{t-\epsilon}\phi)(y) - P_\epsilon(P_{t-\epsilon}\phi)(x)$$

$$= o(\epsilon) + o(\|x - y\|)$$

$$= \mathrm{Err(non-uniqueness)} + \frac{1}{\epsilon}\mathrm{Err}(x - y).$$

The conclusion follows by optimizing in ϵ.

Theorem 10.3.4. *For any Markov family $x \mapsto \mathbb{P}_x$, the map $x \mapsto P(t, x, \cdot)$ is continuous in total variation when $x \in D(A)$.*

The restriction $x \in D(A)$ in the above theorem is due to the fact that we need to ensure the existence of smooth solutions. It can be lowered to $x \in D(A^{1/4+})$, that is, up to the critical space (see Romito, 2011a).

10.3.4 Some remarks on uniqueness

For stochastic (partial) differential equations we may have different notions of uniqueness, regardless of the model we are studying (namely, without introducing any criterion originating from the physics of the problem, such as entropy solutions, etc.). On the one hand there is the notion of path-wise uniqueness, which corresponds to the standard uniqueness for ODE/PDE. On the other hand we may ask for a weaker statement, weak uniqueness, that is uniqueness of distributions.

Noise might be a promising crucial ingredient for uniqueness; see Flandoli, Gubinelli, & Priola (2010). A wider and deeper discussion can be found in Flandoli (2011). Here we only point out two simple (and ineffective so far) criteria for uniqueness and regularity due to Flandoli & Romito (2008):

- If for some (smooth) initial condition there is a smooth solution on a (possibly small but deterministic) time interval, then the problem is well posed.
- If for some initial condition uniqueness in law holds on a (possibly small but deterministic) time interval, then uniqueness in law holds for all initial conditions.

We shall see below in Section 10.4.2 a criterion of uniqueness in terms of invariant measures.

10.3.4.1 Some examples in finite dimension

A standard example of non-uniqueness of an elementary ODE is the equation $\dot{x} = \sqrt{|x|}$ (that we have examined in Example 10.3.3 and we will see again in Example 10.4.4). It is well-known that by adding a Gaussian perturbation $\mathrm{d}x = \sqrt{|x|}\,\mathrm{d}t + \mathrm{d}B$ uniqueness (path-wise) is restored. This is part of a general phenomenon, see for instance Stroock & Varadhan (1979) and Krylov & Röckner (2005). Notice that we would also restore uniqueness by adding something of order one, say $\dot{x} = \sqrt{|x|}+ 1$, the difference is that noise is zero plus random fluctuations.

Weak uniqueness can hold without path-wise uniqueness, as in the Tanaka equation $\mathrm{d}x = \mathrm{sgn}\,(x)\,\mathrm{d}B$. Here all solutions are Brownian motions, hence they all have the same distributions, but there is no pathwise uniqueness since, for instance, if x is a solution, then so is $-x$. To have examples of non–uniqueness of distributions we need to allow degeneracy in the noise coefficient, for instance as in the Girsanov equation $\mathrm{d}x = |x|^\alpha\,\mathrm{d}B$, with $\alpha < \frac{1}{2}$. This problem has an infinite-dimensional counterpart, where several interesting phenomena happen; see Burdzy, Mueller, & Perkins (2010), Mueller, Mytnik, & Edwin (2012) for more details.

Anyway, in dimension $d = 1$ there is a rather complete understanding Engelbert & Schmidt (1985), and the Girsanov example describes a quite universal picture.

10.3.4.2 About uniqueness of the martingale problem

A way to understand uniqueness of distributions is to understand the generator of the process solution of (10.1). Formally, we expect that the generator is

$$\mathscr{L} = \frac{1}{2}\,\mathrm{Tr}[\mathcal{S}^\star \mathcal{S}D^2] - \langle -\nu\Pi_L\Delta u + \Pi_L\big((u \cdot \nabla)u\big), D\rangle,$$

where $\mathcal{S}$ is the operator colouring the noise and Π_L is the Leray projector. It turns out that each of the Markov solutions discussed in Section 10.3.2 is the unique solution of the so-called martingale problem associated to a suitable generator, as stated in the next theorem.

Theorem 10.3.5 (Romito, 2011b). *Given a Markov solution $(\mathbb{P}_x)_{x\in H}$, there exists a unique closed linear operator*

$$\mathscr{L} : D(\mathscr{L}) \subset C_b(D(A)) \to C_b(D(A))$$

such that for all $\lambda > 0$ *and* $\varphi \in C_b(D(A))$,

$$R_\lambda(\mathscr{L})\varphi(x) = \int_0^\infty e^{-\lambda t} P_t \varphi(x) \, dt,$$

where $(P_t)_{t \geq 0}$ *is the transition semigroup associated to the given Markov solution, and* $R_\lambda(\mathscr{L})$ *is the resolvent of* $\mathscr{L}$.

The previous theorem holds under the same assumptions on the covariance as in Section 10.3.3. Similar conclusions can be drawn under the assumptions discussed in Romito & Xu (2011) and Romito (2011a).

The problem here is that each operator $\mathscr{L}$ is equal to $\overline{\mathscr{L}}$ on a class of test functions (smooth functions depending on a finite number of Fourier components). This class of functions unfortunately is not sufficient to characterize the operator[1]. Preliminary computations show that an improved knowledge of the tails of the explosion time τ_∞ (see also the next Section) might be promising. We refer to Romito (2011b) and Da Prato & Debussche (2008) for further details.

10.3.5 Some remarks on blow-up

The aim of this section is to give a brief overview on blow-up and which kind of noise might be more effective to delay or even prevent the emergence of singularities. To our knowledge, Flandoli & Romito (2002a) (see also the related worksby Flandoli & Romito, 2001, 2002b and Romito, 2006) is the only work concerned with singularities for the Navier–Stokes with random perturbations.

Since so far we do not know if the Navier–Stokes equations develop a singularity, it is meaningful to consider simpler models, such as the one we discuss in Section 10.3.5.3, that keep some of the crucial characteristic of the problem (10.1) we are interested in. A recent result of Tao (2016) shows that the analysis of these models may rigorously shed light on the problem of blow-up of (10.1).

The results detailed below (from Romito, 2014e), show that no additive noise can be expected to prevent the formation of singularities. Recent results of Flandoli et al. (2010) and Flandoli (2011) show that a careful choice of the coefficients in the case of state dependent noise might be more promising.

---

[1]To gain intuition, think of the Poisson equation on a bounded domain with two different boundary conditions. Smooth test functions with compact support on the interior of the domain cannot tell the two boundary conditions apart.

10.3.5.1 *The drift matters*

Here we focus on additive noise and we wish to understand if it may have (and possibly how) a significant effect in preventing singularities. As we shall see, the situation is deeply different with respect to the problem of uniqueness. The effect of noise is more related to the stability of blow-up. It may even happen that noise creates singularities when there are none without noise. Consider the problem $\dot{x} = x^2 \sin x$. Clearly there is a global bounded solution for every initial condition, and it is not difficult to see that when adding noise, solutions blow-up[2]. In two dimensions both cases may happen (see Scheutzow, 1993), namely there are two suitable smooth fields $b : \mathbb{R}^2 \to \mathbb{R}^2$ such that if one consider the ODE $\dot{x} = b(x)$ and the SDE $\mathrm{d}x = b(x)\,\mathrm{d}t + \mathrm{d}B$ then

- the ODE explodes for all initial conditions, the SDE has global solutions for all initial conditions with probability one, and there is even an invariant measure,
- the ODE is non-explosive for all initial conditions and the point $(0,0)$ is asymptotically stable, the SDE explodes with positive probability.

Stability is also related to the probability of blow-up, namely if blow-up happens with positive probability or with probability one. Let us consider the equation $\mathrm{d}x = b(x)\,\mathrm{d}t + \mathrm{d}B$, where b is one of the two functions

$$b_\pm(x) = \begin{cases} x^2 & x \geq 0, \\ \pm x & x < 0. \end{cases}$$

When the drift is b_+, blow-up happens with positive probability. When the drift is b_-, blow-up happens almost surely. In view of next sections, we notice that in both cases

- there are T_0, p_0 and a closed set B with *open interior* such that for all initial conditions in B, $\mathbb{P}[\tau_\infty \leq T_0] \geq p_0$,
- the blow-up happens only on the positive 'side': for all $p \geq 1$, namely $E[\sup_t |x_-|^p] < \infty$.

10.3.5.2 *A criterion for the a.s. blow-up*

Define the blow-up time τ_∞ of a stochastic equation (in finite or infinite dimension)

$$\mathrm{d}x = b(x)\,\mathrm{d}t + \mathrm{d}B$$

[2]The role of the noise here is to help overcome the barriers created by the zeroes of sin.

as $\tau_\infty = \sup_n \tau_n$ and

$$\tau_n = \inf\{t : H(x_t) \geq n\}$$

for some quantity of interest for the problem. For instance, $H(x) = |x|$ in finite dimension, H is some norm in a smaller space for stochastic PDEs. Define

$$\flat(t, x_0) = \mathbb{P}_{x_0}[\tau_\infty > t].$$

Then clearly $\flat(0, x_0) = 1$, $\flat(\cdot, x_0)$ is non-increasing, and $t \mapsto \flat(t, x_0)$ is continuous in t (up to technical details). Set

$$\flat(x_0) = \lim_{t \to \infty} \flat(t, x_0) = \inf_t \flat(t, x_0) = \mathbb{P}_{x_0}[\tau_\infty = \infty].$$

In general we cannot claim that $\flat(x_0) \in \{0, 1\}$, as seen in the examples of previous section. On the other hand a 0–1 law still holds for a the supremum of these probabilities.

Theorem 10.3.6 (0–1 law for explosions). *Let $\flat$ be as above. Then* $\sup_{x_0} \flat(x_0) \in \{0, 1\}$.

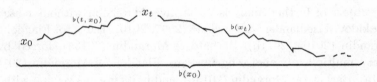

Fig. 10.3. Proof of Theorem 10.3.6.

The proof of the above theorem is sketched in Figure 10.3. The idea now is that if we can prove an upper bound for $\flat$ that keeps $\flat$ away from 1, then by the 0–1 law, $\flat \equiv 0$.

The idea for the upper bound is based on the stability of blow-up and conditional recurrence. Assume there are a closed set B_∞, $p_0 \in (0, 1)$ and $T_0 > 0$ such that

- $\mathbb{P}_y[\tau_\infty \leq T_0] \geq p_0$ for every $y \in B_\infty$,
- $\mathbb{P}_x[\tau^{B_\infty} = \infty | \tau_\infty = \infty] = 0$ for every x,

where τ^B is the hitting time of B, then $\flat(x) = \mathbb{P}_x[\tau_\infty = \infty] \leq \frac{1}{1+p_0}$, hence $\flat \equiv 0$. The heuristic idea here is that the process keeps coming back to the set where blow-up is likely, and once there it tries to blow up, see Figure 10.4. One can look at these trials as coin tossing. Sooner or later both sides will show up. The assumptions above are needed to have a uniform control of the bias of the coin.

Fig. 10.4. Conditional recurrence.

10.3.5.3 The dyadic model of turbulence

In this section, following Romito (2014e), we apply the general criterion explained before to the viscous dyadic model driven by additive noise.

The dyadic model was introduced by Katz & Pavlović (2005) as a simplification of the motion of energy among modes in the Euler equations studied in Friedlander & Pavlović (2004). The model has been the subject of further analysis in its inviscid version without noise by Cheskidov, Friedlander, & Pavlović (2007, 2010), Barbato, Flandoli, & Morandin (2010a, 2011b), Barbato & Morandin (2013a), forced by a special multiplicative noise by Barbato, Flandoli, & Morandin (2010b, 2011a), Barbato & Morandin (2013b), and in its viscous version without noise by Cheskidov (2008), Cheskidov & Friedlander (2009), Barbato, Morandin, & Romito (2011c, 2014) and with noise Romito (2014e). A generalized version on trees, closer to the formulation of Friedlander & Pavlović (2004), has been studied in Barbato et al. (2013) and Bianchi (2013).

A simple derivation of the model by Katz & Pavlović (2005) is as follows. The idea is to look at a solution u on $\mathbb{R}^3$ of the Euler equations and write a simplified version of the interaction of the energy packets. Consider the dyadic cubes: cubes of size 2^ℓ with vertices on $2^\ell \mathbb{Z}^3$. For a dyadic cube Q, let $\widetilde{Q}$ be the parent cube and $\widehat{Q}$ the children cubes (see Figure 10.5). Write the wavelet expansion of u, for an orthonormal basis $(\omega_Q)_Q$ of L^2, based on the dyadic cubes. If $u = \sum_Q u_Q \omega_Q$, the Euler nonlinearity reads $(u \cdot \nabla)u = \sum_Q 2^{j(Q')} u_Q u_{Q'} \omega_Q \omega_{Q'}$. If only nearest neighbours interactions are kept,

$$[(u \cdot \nabla)u]_Q \approx B_{\text{up}}(u,u) - B_{\text{down}}(u,u) = 2^{\beta(j(Q)+1)} u_Q \sum u_{\widehat{Q}} - 2^{\beta j(Q)} u_{\widetilde{Q}}^2,$$

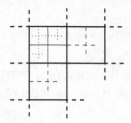

Fig. 10.5. The hierarchy of dyadic cubes.

where 'up' and 'down' refer to the direction of the flow of energy. The value $\beta = \frac{5}{2}$ corresponds, by scaling, to Euler/Navier–Stokes.

A further simplification is achieved by changing the set of indices from dyadic cubes to integers. By adding a viscous dissipation, we finally obtain the *viscous dyadic model*,

$$\dot{x}_n = -\nu\lambda_n^2 x_n + \lambda_{n-1}^\beta x_{n-1}^2 - \lambda_n^\beta x_n x_{n+1} + \sigma_n \dot{w}_n, \qquad n \geq 1,$$

where β measures the relative strength of the dissipative *vs* convective parts. The crucial characteristics preserved by this simplification is formal balance of energy, namely

$$\frac{\mathrm{d}}{\mathrm{d}t}x_n^2 + 2\nu\lambda_n^2 x_n^2 = \underbrace{\lambda_{n-1}^\beta x_{n-1}^2 x_n - \lambda_n^\beta x_n^2 x_{n+1}}_{\text{telescopic!}}$$

$$\rightsquigarrow \quad \frac{\mathrm{d}}{\mathrm{d}t}\sum_{n=1}^\infty x_n^2 + 2\nu\sum_{n=1}^\infty \lambda_n^2 x_n^2 = 0.$$

We summarise some known results, first in the case without noise (when $\sigma_n \equiv 0$ for all n); see Cheskidov (2008), Barbato et al. (2011c),

- positive initial conditions produce positive solutions.
- if $\beta \leq 2$, there is well-posedness (2DNSe–regime),
- if $\beta \leq 3$, uniqueness for positive initial states,
- if $\beta \leq \beta_c \leq 3$ smoothness for positive initial state,
- if $\beta > 3$, blow-up for large enough positive initial state.

Here $\beta_c > \frac{5}{2}$, and so the Navier–Stokes-like case is included. With noise, we know Romito (2014e) that,

- positivity not preserved,
- if $\beta \leq 2$ trivial well-posedness (2DNSe regime),
- if $\beta \leq 3$, path-wise uniqueness,

- if $\beta \leq \beta_c$, smoothness,
- if $\beta > 3$ and $\{n : \sigma_n \neq 0\} \neq \emptyset$, then there is a.s. blow-up from any initial state.

The natural state space of the problem is $\ell^2(\mathbb{R})$; we can consider spaces of regularity (understood as decay, as in Sobolev spaces with respect to Fourier transform),

$$h_\alpha = \big\{ x \in \ell^2 : \|x\|_\alpha^2 := \sum_n (\lambda_n^\alpha x_n)^2 < \infty \big\},$$

and a smooth solution is a solution in h_α (for those values of α compatible with the regularity of the driving noise).

Let us first understand the mechanism that creates blow-up in the case without noise (see Cheskidov, 2008). Assume $\beta > 3$. There is explosion if the initial condition is *positive* and *large enough* in h_α for some suitable $\alpha > 0$. Indeed,

$$\frac{\mathrm{d}}{\mathrm{d}t}\|x\|_\alpha^2 + 2\nu\|x\|_{\alpha+1}^2 \approx \sum_n \lambda_n^{\beta+2\alpha} x_n^3 \approx c\|x\|_{\alpha+1}^3, \qquad (10.5)$$

where the positivity of the solution makes the '$\approx$' rigorous. Hence

$$\frac{\mathrm{d}}{\mathrm{d}t}\|x\|_\alpha^2 \gtrsim c\|x\|_{\alpha+1}^3 - 2\nu\|x\|_{\alpha+1}^2 \gtrsim c\|x\|_\alpha^3, \qquad (10.6)$$

if $\|x(0)\|_\alpha$ is large enough.

The critical exponent for the dyadic model is $\beta - 2$, and local 'regular' solutions exist with initial condition in h_α, with $\alpha > \beta - 2$. Indeed, this number is the decay rate of solutions which yields the equilibrium between the linear and the non–linear part:

$$\lambda_n^2 x_n \approx \lambda_{n-1}^\beta x_{n-1}^2 - \lambda_n^\beta x_n x_{n+1} \approx \lambda_n^\beta x_n^2,$$

that is $x_n \approx \lambda_n^{2-\beta}$.

When we have at least one component forced by noise, blow-up is almost sure.

Theorem 10.3.7. *Assume $\sigma_n \neq 0$ for at least one index $n \geq 1$. For every $x \in h_\alpha$, $\alpha > \beta - 2$, and every martingale solution starting at x, $\mathbb{P}_x[\tau_\infty < \infty] = 1$.*

The proof of this theorem is based on the criterion explained in Section 10.3.5.2. The strategy is to consider a perturbation of the deterministic estimates (10.5) and (10.6) in order to find a good set B_∞ where the probability of blow-up is bounded from below. The main difficulty in identifying B_∞ is that positive states are a 'thin' set in ℓ^2. On the other hand, when seen at an appropriate scale, the noise perturbation is small, so that, if not positive, solutions are still not too negative. This identifies B_∞ as a subset of states that are 'quasi-positive' and large is some suitable norm.

To prove the conditional recurrence, we prove that it is implied by recurrence of balls in ℓ^2 (which is true by the standard energy estimate). Indeed, fix a ball of radius M in ℓ^2, then with a probability p_2 there

is contraction of negative components (so that the solution is 'quasi-positive'), while keeping the size in ℓ^2 not much larger than M. With probability p_1 it is possible to keep the negative components small, while making the positive part of the first noisy component large. The two steps together have lead the system into B_∞. The events that force the system the way we want depend on the outcomes of the noise, and so are mutually independent. Hence recurrence of B_∞ is reduced to recurrence of balls in ℓ^2.

10.3.6 Existence of densities for finite dimensional functionals

There are several reasons to be interested in densities for the solutions of (10.1).

First of all, when dealing with a stochastic evolution PDE, the solution depends not only on the time and space independent variables, but also on the 'chance' variable, and the existence of a density for the law of (some functionals of) the solution is thus a form of regularity with respect to this new variable.

We will be particularly interested in densities for finite dimensional (spectral) projections of the solution. By the results in Section 10.3.2, it is sufficient to show that the laws of two solutions agree if they agree at each time. To show that the laws at each time coincide it is sufficient to show that the finite-dimensional projections are the same. Thus, the analysis of densities can be a first step towards a proof of uniqueness of the distributions[1].

An alternative proof of uniqueness, involving the invariant measures, is

---

[1]Unfortunately, we are not able to proceed beyond this first step so far.

presented in Theorem 10.4.2. It is interesting, although not unexpected, that the densities of stationary processes (see Section 10.4) are smoother than the densities of any other solution (see Debussche & Romito, 2014), so that the strategy outlined above might proceed further when we consider the additional smoothing of stationary solutions together with the aforementioned Theorem 10.4.2.

When looking for densities, we face two non-trivial problems. The first concerns the problem of a reference measure for densities. One reason to consider finite-dimensional functionals is that there is no canonical reference measure in infinite dimension. Any choice should be necessarily tailored to the problem at hand, and in our case we do not know enough of the problem (10.1) for this purpose (see Romito (2013) for more details in this direction, see Section 10.4.3 for some candidates).

The second problem is related to the difficulty in proving regularity and uniqueness (either with or without noise). Indeed, to show existence of densities for solutions of stochastic equations a fundamental and classical tool is the Malliavin calculus, a differential calculus where the differentiating variable is the underlying noise driving the system. The Malliavin derivative $\mathcal{D}_H u(t)$, the derivative with respect to the variations of the noise perturbation, is given as

$$\mathcal{D}_H u = \lim_{\epsilon \downarrow 0} \frac{u(W + \epsilon \int H \, ds) - u(W)}{\epsilon},$$

where we have written the solution u as $u(W)$ to show the explicit dependence of u from the noise forcing. We point, for instance, to Nualart (2006) for further details and definitions. In dimension two the general program proceeds (see Mattingly & Pardoux, 2006) by providing smooth densities for the finite-dimensional projections of the solution. When we turn to three dimensions, we notice that the Malliavin derivative $\mathcal{D}_H u$ of the solution u of (10.1), as a variation, satisfies the linearization around the solution, namely,

$$\frac{d}{dt}\mathcal{D}_H u - \nu \Delta \mathcal{D}_H u + (u \cdot \nabla)\mathcal{D}_H u + ((\mathcal{D}_H u) \cdot \nabla)u = SH,$$

and good estimates on $\mathcal{D}_H u(t)$ originate only from good estimates on the linearization of (10.1), which are not available so far. This settles the need for methods to prove existence and regularity of the density that do not rely on this calculus, as done in Debussche & Romito (2014). For other works in this direction, see for instance De Marco (2011), Bally & Caramellino (2012), Kohatsu-Higa & Tanaka (2012) and Hayashi, Kohatsu-Higa, & Yûki (2013a,b).

10.3.7 Besov bounds for the densities

There are several possible strategies to prove the existence of the density of a random variable. We have already been convinced that Malliavin calculus does not work. In Debussche & Romito (2014) three different strategies are presented. The first strategy is based on the idea introduced by Flandoli & Romito (2007) that, under suitable assumptions on the driving noise, Markov solutions have laws that are absolutely continuous with respect to the laws of the local strong solutions. As observed in Romito (2014a) (see also Romito, 2014c), the validity of this observation, when properly reformulated, goes beyond Markov solutions. A second strategy is based on the Girsanov transformation and, as the previous strategy, provides only a qualitative result, namely the existence of densities, without any further regularity property.

The third strategy, that we are going to briefly detail below, yields regularity of the density in terms of Besov spaces. Let us recall the definition of Besov spaces. The general definition is based on the Littlewood–Paley decomposition, but it is not the best suited for our purposes. We shall use an alternative equivalent definition (see Triebel, 1983, 1992) in terms of differences. Define

$$(\Delta_h^1 f)(x) = f(x+h) - f(x), \qquad (\Delta_h^n f)(x) = \Delta_h^1 (\Delta_h^{n-1} f)(x)$$

then the following norms, for $s > 0$, $1 \le p \le \infty$, $1 \le q \le \infty$,

$$\|f\|_{B_{p,q}^s} = \|f\|_{L^p} + \left(\int_{\{|h| \le 1\}} \frac{\|\Delta_h^n f\|_{L^p}^q}{|h|^{sq}} \frac{\mathrm{d}h}{|h|^d} \right)^{\frac{1}{q}},$$

(with obvious modifications when $q = \infty$), where n is any integer such that $s < n$, are equivalent norms of $B_{p,q}^s(\mathbb{R}^d)$ for the given range of parameters.

The technique introduced by Debussche & Romito (2014) is based on the following analytic lemma, which provides a quantitative integration by parts. The lemma is implicitly given in Debussche & Romito (2014) and explicitly stated in Romito (2015).

Lemma 10.3.8 (smoothing lemma). *Let μ be a finite measure on $\mathbb{R}^d$, $m \ge 1$ an integer, $0 < s \in \mathbb{R}$, $\alpha \in (0,1)$, with $\alpha < s < m$, and a constant $c_{11} > 0$ such that for every $\phi \in C_b^\alpha(\mathbb{R}^d)$ and $h \in \mathbb{R}^d$,*

$$\left| \int_{\mathbb{R}^d} \Delta_h^m \phi(x) \, \mu(\mathrm{d}x) \right| \le c_{11} |h|^s \|\phi\|_{C_b^\alpha},$$

then μ has a density f_μ with respect to the Lebesgue measure on $\mathbb{R}^d$ and

$f_\mu \in B_{1,\infty}^r$ *for every* $r < s - \alpha$. *Moreover, there is* $c_{12} = c_{12}(r)$ *such that*

$$\|f_\mu\|_{B_{1,\infty}^r} \leq c_{11}c_{12}.$$

The idea is the same used with Malliavin calculus: there is a smoothing effect (that is captured by the above lemma) and this must originate from the random perturbation. We use the random perturbation to perform the 'fractional' integration by parts along the noise to be used in the above lemma. The bulk of this idea can be found in Fournier & Printems (2010). Our method is based on the one hand on the idea that the Navier–Stokes dynamics is 'good' for short times, and on the other hand that Gaussian processes have smooth densities. When trying to estimate the Besov norm of the density, we approximate the solution by splitting the time interval in two parts.

On the first part the approximate solution u_ϵ is the same as the original solution, on the second part the nonlinearity is killed. By Gaussianity it is enough to estimate the increments of the density of u_ϵ. Since u_ϵ is the one-step explicit Euler approximation of u, the error in replacing u by u_ϵ can be estimated in terms of ϵ. By optimizing the increment versus ϵ we have an estimate on the derivatives of the density.

The regularity of the density can be slightly improved from $B_{1,\infty}^{1-}$ to $B_{1,\infty}^{2-}$ if u is the stationary solution, namely the solution whose statistics are independent from time (see Section 10.4)

Proposition 10.3.9. *Given* $x \in H$ *and a finite-dimensional subspace* F *of* $D(A)$ *generated by the eigenvectors of* A, *i.e.* $F = \mathrm{span}[e_{n_1}, \dots, e_{n_F}]$ *for some arbitrary indices* $n_1, \dots, n_F$, *assume that* $\pi_F S$ *is invertible on* F. *Then for every* $t > 0$ *the projection* $\pi_F u(t)$ *has an almost-everywhere positive density* $f_{F,t}$ *with respect to the Lebesgue measure on* F, *where* u *is any solution of* (10.3), *which is limit point of the spectral Galerkin approximations.*

Moreover, for every $\alpha \in (0,1)$, $f_{F,t} \in B_{1,\infty}^\alpha(\mathbb{R}^d)$ *and for every small* $\epsilon > 0$, *there exists* $c_{13} = c_{13}(\epsilon) > 0$ *such that*

$$\|f_{F,t}\|_{B_{1,\infty}^\alpha} \leq \frac{c_{13}}{(1 \wedge t)^{\alpha+\epsilon}}(1 + \|x\|_H^2).$$

The time regularity of the density can be also investigated using similar ideas. Since we are studying a stochastic evolution, it is reasonably expected that there is regularity in time and that this is 'half' the regularity in space (Brownian scaling). This is confirmed by the following result.

Proposition 10.3.10 (Romito, 2015). *Under the same assumptions of Proposition 10.3.9, for every* $\alpha \in (0,1)$ *there exists* $c_{14} > 0$ *such that for every* $s, t > 0$,

$$\|f_{F,t} - f_{F,s}\|_{B_{1,\infty}^\alpha} \le c_{14}\Big(\sup_{r \in [s,t]} \|f_{F,r}\|_{B_{1,\infty}^\alpha} \Big)\big)|t - s|^{\frac{\alpha}{2}}.$$

The above result is based again on the splitting idea we have explained. Unfortunately the splitting works only when one can exploit a sort of integration by parts. This is not possible when doing an estimate of the densities in L^1. To this purpose, in Romito (2015), the L^1 estimate of the time increments follows from (an appropriate version of) the Girsanov change of measure.

Besov bounds work well with finite-dimensional projections because they allow us to avoid the difficulty of low regularity of solutions of Navier–Stokes in three dimension. On the other hand the method based on Markov solutions works well, at least qualitatively, for any finite-dimensional functional. It may be interesting to provide some 'ad-hoc' strategy to get regularity of densities of some quantities of interest for the equations. In Romito (2014b) there is a proof of regularity of densities for the energies $\mathcal{E}_\alpha(t, u) = \|u(t)\|_\alpha^2$, for negative α. More precisely, the following result holds.

Proposition 10.3.11 (Romito, 2014b). *Given a weak martingale solution* u *of* (10.1), *and a number* $\alpha > \frac{3}{4}$, *the real-valued random variable* $\mathcal{E}_{-\alpha}(t, u)$ *has a density* $f_{t,\alpha}$ *with respect to the Lebesgue measure on* $\mathbb{R}$, *for every* $t > 0$. *Moreover,*

- $f_{t,\alpha} \in B_{1,\infty}^r(\mathbb{R})$ *for every* $r < 2\alpha - \frac{3}{2}$ *if* $\alpha < \frac{5}{4}$,
- $f_{t,\alpha} \in B_{1,\infty}^r(\mathbb{R})$ *for every* $r < 1$ *if* $\alpha \ge \frac{5}{4}$.

A similar statement can be also obtained for the α-dissipation case $\mathcal{D}_\alpha(t, u) = \int_0^t \|u(s)\|_{1+\alpha}^2 \, ds$ (again with α negative), as well as for the fundamental energy balance $\mathcal{E}_0(t, u) + 2\nu \mathcal{D}_0(t, u)$, namely,

$$\|u(t)\|_H^2 + 2\nu \int_0^t \|u(s)\|_V^2 \, ds.$$

In the latter case, the key point is that we can exploit the cancellation property of the Navier–Stokes nonlinearity. For α negative but larger than the threshold of Proposition 10.3.11, the contribution of the nonlinearity has no good control.

10.3.8 Additional remarks

An interesting question, that has been completely answered for the two-dimensional case by Mattingly & Pardoux (2006), concerns the existence of densities when the covariance of the driving noise is essentially non-invertible. The typical perturbation in (10.1) we consider here is

$$\dot{\eta}(t,y) = \sum_{k \in \mathcal{K}} \sigma_{\mathbf{k}} \dot{\beta}_{\mathbf{k}}(t) e_{\mathbf{k}}(y),$$

where $\mathcal{Z} \neq \mathbb{Z}_\star^3$ and is usually much smaller (finite, for instance). The idea is that the noise influence is spread, by the nonlinearity, to all Fourier components. The condition that should ensure this has been already well understood by Romito (2004), and corresponds to the fundamental algebraic property that $\mathcal{K}$ should generate the whole group $\mathbb{Z}^3$.

It is clear that the method we have used to obtain Besov bounds cannot work in this case, because the nonlinearity plays a major role. On the other hand in Romito (2014a) we prove, using ideas similar to those used in the strategy with Markov solutions, the existence of a density. No regularity properties are possible, though. See also Romito (2005), Romito & Xu (2011) for other relevant results on the Navier–Stokes equations in dimension three with 'hypoelliptic noise'.

Another issue that is generically applicable to any statement related to the Navier–Stokes equations in 3D, is the universality of the result obtained. Since we do not know if there is a unique distribution, it may be possible that the densities of solutions obtained by different methods may have different properties. In a way this is reminiscent of the problem of suitable weak solutions introduced by Scheffer (1977). Only much later it has been proved that solutions obtained by the spectral Galerkin methods are suitable (under some non-trivial conditions though; see Guermond, 2006), and hence results of partial regularity are true for those solutions. In Romito (2014c) we establish a 'transfer principle' that, roughly speaking, states that as long as we can prove existence of a density for a finite-dimensional functional of the solution and for a weak solution that satisfies *weak-strong uniqueness*, then existence of a density holds for any other solution satisfying weak-strong uniqueness and a closure property with respect to conditional probabilities.

In other words, by the transfer principle, we can prove existence of a density for solutions obtained from Galerkin approximation, and this result will extend straight away to any other solutions, for instance those produced by the Leray regularization (see for instance Lions, 1996). We can use the special properties of Markov solutions given in Flandoli & Romito (2008) and Romito (2010a) to prove existence of densities for a large class of finite dimensional functionals, as done in the first part of Debussche & Romito (2014); again this extends immediately to any solution.

An important limitation of the transfer principle is that it applies only on quantities depending only on one time. This for instance rules out the results of Proposition 10.3.10. Moreover, the transfer principle is qualitative in nature, as it may transfer only the existence, and in general no quantitative information can be inherited. Nevertheless, in the case of stationary solutions, the transfer principle (with some loss) can be also made quantitative.

10.4 Invariant measures

Existence of invariant measures for stochastic equations is a classical topic; we refer to Da Prato & Zabczyk (1996) for details. In fact, it dates back to original ideas of Kolmogorov, to add noise to an equation to find a unique invariant measure (something that is way more difficult when randomness is in the initial condition, see Section 10.2.5), and then to study the zero-noise limit to 'select' the most interesting invariant measure of the noise-less equation.

Thus the most interesting part of the analysis in this subject concerns the uniqueness of invariant measures, and possibly related ergodic theorems. The theory in the two-dimensional case, starting from the first results of Flandoli & Maslowski (1995), is well developed, both on the side of rough noise (see for instance Da Prato & Debussche, 2002), and on the side of smooth finite-dimensional noise (in some way, the most physical one when turning to turbulence); see Hairer & Mattingly (2006).

In dimension three the theory is less developed, and so far we only know that some special solutions (the Markov solutions of Da Prato & Debussche (2003a) or Flandoli & Romito (2008) discussed in Section 10.3.2) admit an invariant measure which is 'unique'. This uniqueness statement has to be properly and carefully understood. The main limitation in these results is in noise. Indeed, these results require that the noise acts on all modes and that the decay of noise coefficients is controlled from above and below (with a minimum of flexibility in the control, see Romito, 2011a).

10.4.1 Existence and uniqueness of invariant measures in 3D

The fundamental issue when dealing with invariant states in the three-dimensional case, hence in the case where the dynamics is not well defined, is to identify a good definition of an invariant state. A fairly general definition is a *stationary solution*. Consider as a state space the set of all trajectories, for instance $\Omega = C([0, \infty); \mathcal{V}')$ (and further requirements, if necessary in order to get compactness), where $H \subset \mathcal{V}'$ is large enough so that solutions are continuous. A stationary solution is a probability measure on the state space such that trajectories are solutions of (10.1) (with or without noise) and such that the measure is invariant with respect to the time-shift, namely with respect to the maps $\tau_t : \Omega \to \omega$ defined as

$$\tau_t(\omega)(s) = \omega(t + s), \qquad s \geq 0.$$

Roughly speaking, the action of τ_t is to cut out the first part of the trajectory, thus a stationary solution depends only on the 'tails' of the trajectories. The time marginal of a stationary solution is the candidate invariant measure, although there is not a well-defined flow with respect to which the measure is invariant.

The techniques of Da Prato & Debussche (2003a), Flandoli & Romito (2008) are specially tailored to define a flow of solutions of (10.1). In this setting invariant measures are a meaningful notion. As already mentioned, existence of an invariant measure is straightforward, after all the dynamics without forcing has the zero as the unique globally stable point, at least in H. Compactness concludes the argument.

Uniqueness is obtained in Da Prato & Debussche (2003a) as a by-product (through Doob's theorem, see Da Prato & Zabczyk, 1996) of two fundamental properties: continuity of transition probabilities, discussed in Section 10.3.3, and the full support. Both properties rely on the strong assumptions on the noise we have mentioned.

Regarding convergence to the unique invariant measure, we know that it is exponentially fast Odasso (2007), Romito (2008). It is worth mentioning that in Odasso (2007) convergence of a solution towards a stationary solution is obtained for every limit probability of Galerkin approximations. In Romito (2008) convergence holds only for Markov solutions.

A completely different approach is presented in Bakhtin (2006), based on ideas we will see later in Section 10.5.1.3, but is restricted to bounded (and small enough) noise, although it ensures uniqueness of a stationary solution.

In conclusion, state-of-the-art results in this setting are still very far from the strong results obtained in the two-dimensional case in Hairer & Mattingly (2006). A mild relaxation on the non-degeneracy of the noise has been given in Romito & Xu (2011).

10.4.2 Uniqueness criterion through invariant measures

We have already seen that, under suitable assumptions on the driving noise, every Markov solution has a unique invariant measure. As in principle there can be several different Markov solutions, so can invariant measures.

Here we wish to discuss the different notions of invariant measures we can consider and a uniqueness criterion for the law of (10.1) based on invariant measures.

Define a stationary solution $\mathbb{P}_\star$ as a probability measure on Ω_{NS} that is invariant for the time shift on the path space. There are several ways to ensure existence of stationary solutions, either without noise (see Flandoli & Romito, 2001), or with noise (see Romito, 2010b). One way is also provided by Markov solutions since if $(\mathbb{P}_x)_{x \in H}$ any such solution and $\mu_\star$ is its invariant measure, then $\mathbb{P}_\star = \int \mathbb{P}_x \, \mu_\star(\mathrm{d}x)$ is a stationary solution.

Define the following sets

$$\mathscr{I} = \{\mu \in \mathrm{Pr}(H) : \mu \text{ is the marginal of a stationary solution}\},$$

$$\mathscr{I}_m = \{\mu \in \mathrm{Pr}(H) : \mu \text{ is the invariant measure of a Markov solution}\},$$

$$\mathscr{I}_e = \{\mu \in \mathscr{I}_m : \mu \text{ maximizes the selection procedure}\},$$

where the selection procedure is the one outlined in Section 10.3.2. It is easy to check (see Romito (2008)) that $\mathscr{I}_e \subset \mathscr{I}_m \subset \mathscr{I}$, and that $\mathscr{I}$ is compact.

The set of invariant measures associated to Markov solutions has a robust structure, so that several uniqueness results and characterizations are possible. For instance, we have the following result due to Romito (2008).

Proposition 10.4.1. *If $\mu_\star \in \mathscr{J}_e$, then the stationary solution $\mathbb{P}_\star$ associated to $\mu_\star$ is the unique stationary measure in $\mathscr{C}(\mu_\star)$.*

If we also assume that the covariance of the driving noise is as in Flandoli & Romito (2008) (or as in Romito, 2011a), so that the dynamics is strong Feller and irreducible in a smaller space $\mathcal{W}$ (tipically $\mathcal{W} = V_\alpha$ for a suitable $\alpha > \frac{1}{2}$), then we have the following result (see also Krylov (2004) for a finite dimensional analogue of this result).

Theorem 10.4.2. *Assume that every Markov selection is strong Feller and fully supported on $\mathcal{W}$. Let $(\mathbb{P}_x)_{x \in H}$ and $(\mathbb{P}'_x)_{x \in H}$ be two Markov solutions, with $(\mathbb{P}_x)_{x \in H}$ maximizer in one of the possible maximisation procedures of the selection. If the two families have the same invariant measure, then they coincide on $\mathcal{W}$.*

In different words, if the set $\mathscr{J}_e$ contains only one invariant measure, then the martingale problem associated to the Navier–Stokes equations (10.1) is well posed. The converse is obvious.

We have not been able to find a way to apply the criterion. On the other hand, under the same assumptions of the above theorem, we know that the measures in $\mathscr{J}_m$ are all equivalent measures (Romito, 2008, Corollary 3.5)

Theorem 10.4.3. *Any μ_1, $\mu_2 \in \mathcal{I}_m$ are equivalent:*

$$\mu_2 = \frac{\mathrm{d}\mu_2}{\mathrm{d}\mu_1} \mu_1.$$

In different words: any *almost-sure* event is universal and the property holds independently of the (Markov) solution. To improve the reliability of the model and give a quantitative measure of uncertainty regarding events related to the fluid motion a better understanding of the density $\frac{\mathrm{d}\mu_2}{\mathrm{d}\mu_1}$ is needed. In principle, by having, for instance, stronger summability of the density than L^1, it may be possible to ensure that

$$\mu_1[A] \approx 1 \quad \Longrightarrow \quad \mu_2[A] \approx 1,$$

so that events of high probability for one solution are strongly likely for any other, independently of the uniqueness of distributions. This in a way justifies the line of research initiated in Sections 10.3.6-10.3.8.

Example 10.4.4 (Romito, 2011b). In order to understand the questions of uniqueness and non-uniqueness in relation with invariant measures, let us consider a variation of the elementary example 10.3.3, $\dot{X} = -X + \sqrt{X}$, with initial condition $X(0) = x \in [0, 1]$. The problem has a unique solution $X_x(\cdot)$ for $x \neq 0$ and the family of solutions $\{X_a^\star = X^\star((t-a) \vee 0) : a \geq 0\}$ for $x = 0$, where $X^\star$ is the unique solution starting at 0 such that $X^\star(t) > 0$ for all $t > 0$.

If $\mathscr{C}(x)$ denotes the set of all solutions starting at x, viewed as probability measures on the path space $C([0, \infty); [0, 1])$, then $\mathscr{C}(x) = \{\delta_{X_x}\}$ for $x \in (0, 1]$, where δ_{X_x} is the Dirac measure concentrated on X_x.

If $x = 0$, a solution starts at 0 and stays for an arbitrary time, then follows the solution $X^\star$ (suitably translated). So the *departing* time from 0 can be interpreted as a random time T whose law can be arbitrary. Therefore any selection of solutions is completely described by a single random variable on $[0, \infty]$. It is easy to see that a selection of solutions

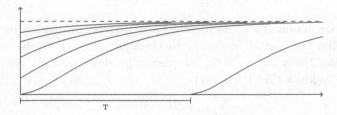

is Markov if and only if T is exponential (the lack of memory plays a major role), including the degenerate cases of infinite or zero rate (namely, $T = 0$ or $T = \infty$).

Denote by $(\mathbb{P}_x^a)_{x \in [0,1]}$ the Markov family with rate a. We shall call *extremal* all those Markov solutions that can be obtained by the selection procedure. It turns out that the only extremal families are those corresponding to $a = 0$ and $a = \infty$.

Concerning invariant measures, we notice that $(\mathbb{P}_x^a)_{x \in [0,1]}$ has a unique invariant measure (which is δ_1) if and only if $a < \infty$. Notice that all selections having δ_1 as their unique invariant measure coincide δ_1-almost surely.

If $a = \infty$, there are infinitely many invariant measures (the convex hull of δ_0 and δ_1). As there is no noise in this example, in general we cannot expect the invariant measures to be equivalent.

10.4.3 Explicit invariant measures

With Section 10.2.5 in mind, let us look if we can find a suitable noise that provides an explicit expression for the invariant measure. Let us look for an invariant measure of Gibbs type, namely $\mu = e^{U(x)}\,dx$, so that it has a density with respect to the (non-existent) infinite-dimensional flat measure.

Formally, the Fokker–Planck equation satisfied by the infinite-dimensional density $U = U(t, x)$, $t > 0$, $x \in H$, is

$$\frac{1}{2}\operatorname{Tr}(\mathcal{S}^\star \mathcal{S} D^2 U) + \sum_n \frac{\partial}{\partial x_n}\big(U(\nu\lambda_n x_n + B_n(x))\big) = 0,$$

where $(\lambda_n)_{n\geq 1}$ are the eigenvalues of the Laplace operator, and $\mathcal{S}^\star \mathcal{S}$ is the covariance of the driving noise, and we assume, for simplicity, that there are numbers $\sigma_n \geq 0$ such that $Se_n = \sigma_n$, that is the covariance is diagonal in the usual real Fourier basis $(e_n)_{n\geq 1}$. We postulate that

$$U = e^{-\beta\sum_n \mu_n x_n^2},$$

that is we assume that the invariant measure is a Gaussian measure. We choose the coefficients $(\mu_n)_n$ to be either $\mu_n = 1$ for every n (in dimension 2 or 3), or $\mu_n = \lambda_n$ for every n (in dimension 2). With both choices we know that (formally)

$$\sum_n \mu_n x_n B_n(x) = 0.$$

With our choice of U, we have that

$$\frac{\partial U}{\partial x_n} = -2\beta\mu_n x_n U, \qquad \frac{\partial^2 U}{\partial x_n^2} = -2\beta\mu_n U + 4\beta^2 \mu_n^2 x_n^2 U,$$

hence, by plugging these formulas in the Fokker–Planck equation and using the formula above for B,

$$U\sum_n(-\beta\sigma_n^2\mu_n + 2\beta^2\mu_n^2\sigma_n^2 x_n^2 - 2\nu\beta\lambda_n\mu_n x_n^2 + \nu\lambda_n) = 0,$$

which yields $\nu\lambda_n - \beta\mu_n\sigma_n^2 = 0$, or

$$\sigma_n^2 = \frac{\nu\lambda_n}{\beta\mu_n}.$$

10.4.3.1 Invariant measures from the enstrophy

Let us consider first dimension two and $U = e^{-\beta S(x)}$, where S is the enstrophy. The above computations suggest that we should choose the identity (on H) as covariance (i.e. the driving noise is white in both space and time). Let us look first at the solution z of the linear problem (10.4). A simple computation shows that $\boldsymbol{E}[\|z(t)\|_H^2] = \infty$, hence $\mathbb{P}[z(t) \notin H] = 1$ by Fernique's theorem (see Bogachev, 1998). Since we cannot expect that the solution to the full Navier–Stokes equations could be any smoother than z (in fact it is not), standard methods do not apply. In Da Prato & Debussche (2002) the equation is understood in terms of a renormalized nonlinearity, where the square of the distribution u is understood as a Wick product, very much the same as it was done in the case of the Φ_3^4 model in the stochastic quantisation of Euclidean quantum field theory (see for instance Da Prato & Debussche, 2003b and Hairer, 2015b).

Here we give a rough idea following Blömker & Romito (2013) (where it has been applied to a fourth order problem). Let z_N be the spectral truncation of z (namely, take only modes $|k| \leq N$, then $B(z_N, z_N)$ is well defined and is a Cauchy sequence in $V_{-\alpha}$, $\alpha > 0$. To see the reason for this, let us check a simpler computation that shows that $B(z_N, z_N)$ is bounded in $V_{-\alpha}$ for $\alpha > 0$. The Cauchy sequence statement follows by similar considerations. Each component of z can be written as $z_k = \frac{k^\perp}{|k|} \zeta_k$, where

$$d\zeta_k + \nu |k|^2 \zeta_k = d\beta_k,$$

where $(\beta_k)_{k \in \mathbb{Z}_*^2}$ are independent complex-valued standard Brownian motion (but for $\beta_k = \beta_{-k}$). We first notice that $\boldsymbol{E}[\zeta_m \zeta_n j] = 0$ if $m \neq \pm n$ by independence. Moreover, $\boldsymbol{E}[\zeta_m^2] = 0$ and $\boldsymbol{E}[|\zeta_m|^2] = (\nu |m|^2)^{-1}$ by a direct computation, hence in the sum below, most of the terms cancel out and

$$\boldsymbol{E}[\|B(z_N, z_N)\|_{-\alpha}^2] = \mathbb{E} \sum_k |k|^{-2\alpha} \left| \sum_{m+n=k} (z_n \cdot k) \pi_k z_m \right|^2$$
$$= \sum_k |k|^{-2\alpha} \sum_{\substack{m_1+n_1=k \\ m_2+n_2=k}} \mathbb{E}[(z_{n_1} \cdot k)(z_{n_2} \cdot k)(\pi_k z_{m_1} \cdot \pi_k z_{m_2})].$$

Consider the expectation $\boldsymbol{E}[(z_{n_1} \cdot k)(z_{n_2} \cdot k)(\pi_k z_{m_1} \cdot \pi_k z_{m_2})]$. It is zero unless $m_1 = m_2$ and $n_1 = n_2$, or $m_1 = n_2$ and $m_2 = n_1$ (the case $m_1 = n_1$, $m_2 = n_2$ can be neglected). In both cases the expectation is bounded by $\nu^{-2} |k|^2 |m|^{-2} |n|^{-2}$ and it is easy to check (see for instance Blömker, Romito, & Tribe, 2007, Lemma 2.3) that the sum

$$\sum_k |k|^{-2\alpha+2} \sum_{m+n=k} \frac{1}{|m|^2 |n|^2} \leq c_{15} \sum_k \frac{\log |k|}{|k|^{2\alpha+2}}$$

converges when $\alpha > 0$. It might be possible that a different approximation of z may lead to a different limit object $B(z, z)$ (see the discussion
in Blömker & Romito, 2013).

Once we know that $B(z, z)$ is well defined, we can look for a solution
$u = z + v$, where v solves

$$\partial_t v + \nu A v + B(v, v) + B(z, v) + B(v, z) = -B(z, z).$$

It turns out that v is slightly better than u (the worst part of u is all in
z), so that the products $B(v, v)$, $B(v, z)$ and $B(z, v)$ make sense without
any further consideration. The above equation can be solved by a fixed
point argument. The passage from local to global works using the fact
that the system under consideration has an explicit invariant measure
(see Da Prato & Debussche (2002) for further details).

The whole idea can be pushed even further, because one can allow
for rougher covariance operators, as long as the equation for v makes
sense, possibly by extending the decomposition of u to new factors, for
instance at second order $u = z + w + v$, with $\partial_t w + \nu A w + B(z, z) = 0$,
and v so that formally the equation for u is the correct one.

Unfortunately, the whole method finds a substantial limit, that we
discuss in the next section.

10.4.3.2 Invariant measures from the energy?

The *naive* approach discussed in the previous section finds robust general versions in the works Gubinelli, Imkeller, & Perkowski (2013) and
Hairer (2015b) (see also Hairer, 2015a, 2014b). A general rule to multiply random distributions was introduced by Gubinelli et al. (2013)
based on the Littlewood–Paley decomposition. The approach of Hairer
(2015b) is more abstract and general, and aims to introduce a new kind
of Taylor expansion, where the basic elements of the expansion can be
tailored to the problem. Both methods have been introduced to deal
with problems, such as the Kardar–Parisi–Zhang equation (see Hairer,
2013), where the noise perturbation is so rough that in principle should
have no meaning. Our previous section suggests the general approach
of writing the solution as the contribution of several terms of growing
regularity, up to the point that the 'last' term is smooth enough so
that the nonlinearity makes sense. In general there is no unique way
to do this and in the most challenging problems (such as KPZ), some
renormalization of these quantities may be required.

The structural limitation of the method is that there might be no gain
of smoothing (or worse). For such *critical* (or super-critical) problems

the problem is completely open. A 'rule of thumb' is suggested in Hairer (2015b) to provide some intuition. To simplify, let us consider our case of interest, namely the Navier–Stokes equations driven by the gradient of space-time white noise. From our formal computations in Section 10.4.3, this is the correct noise to ensure that the Gibbs measure associated to the energy is (formally) invariant.

Space-time white noise is worth $-\frac{d}{2} - 1$ derivatives (time is worth one, each space dimension is worth a half), and we loose another derivative for the gradient. The linear dissipation allows to gain two derivatives. Then we plug the result in the nonlinearity. We may look at the Navier–Stokes nonlinearity as $\mathrm{div}\,(u \otimes u)$, which is better here because we need to take the square of a distribution rather than multiply it by its gradient. The external div is taken care of by integration by parts, making things easier.

By this computation of derivatives, it turns out that in dimension two we are dealing with a critical problem (and the problem is much worse in dimension three). In fact, it is possible that it is already worse in dimension two due to some logarithmic divergence typical of the two dimensional Navier–Stokes nonlinearity.

In conclusion, with the knowledge so far the existence of a Navier–Stokes flow leaving the Gibbs measure from the energy invariant is an open problem.

10.5 Other topics

In this last section we collect some additional topics related to the Navier–Stokes equations and probabilistic tools. We will give some detailed ideas only on the problem of probabilistic representation formulas for the solutions of the *deterministic* Navier–Stokes equations (although all formulas provided may be adapted to deal with random forcing). We conclude by recalling a few works related with the statistics on the Navier–Stokes equations with noise.

10.5.1 Representation formulas

We now focus on the equations (10.1) with no random data and we are interested in finding the 'hidden' stochasticity that can allow to represent the solution by a formula containing random terms.

The representation of the velocity field of a fluid through characteristics is very natural in fluid dynamics, due to the possibility of describing the motion of a fluid using a Lagrangian point of view. On the other hand the idea of using random processes to give a representation of solutions is as old as Brownian motion. The two facts capture two of the main phenomena of the motion of viscous fluids, transport and diffusion. Indeed, in dimension two this is sufficient to obtain probabilistic representation formulas for the solutions (see Busnello, 1999).

It turns out that in dimension three another phenomenon, vortex stretching, has to be taken into account, and this makes representation formulas more challenging. Two versions, Busnello, Flandoli, & Romito (2005) and Constantin & Iyer (2008) detailed below, give a description in these terms.

An alternative definition, that roughly speaking focuses on the interaction of energy among modes can be given by means of branching processes (see Le Jan & Sznitman, 1997). Here we will give a simplified idea following Blömker et al. (2007). The same approach makes it possible to define stationary solutions for the equations driven by bounded (non-Gaussian) stationary noise (see Bakhtin, 2006).

Other probabilistic representation formulas, not presented here have been developed by Esposito et al. (1988), Esposito & Pulvirenti (1989), Albeverio & Belopolskaya (2003) and Cruzeiro & Shamarova (2009).

10.5.1.1 Representation through noisy Lagrangian trajectories

We consider here results by Busnello et al. (2005). Let us start from the equation for the vorticity ($\xi = \text{curl}\, u$),

$$\partial_t \xi + (u \cdot \nabla)\xi = \nu \Delta \xi + D_u\, \xi,$$

where $D_u = \frac{1}{2}(\nabla u + (\nabla u)^T)$ is the deformation tensor, that is the symmetric part of ∇u. The 'stretching' term $D_u\, \xi$ is the responsible for this three-dimensional phenomena and is not present in the equation for the two-dimensional vorticity.

Let us first consider the two-dimensional vorticity of an inviscid fluid. Vorticity is transported along the flow induced by the velocity, so that the quantity $\xi(t, X_t)$ is conserved when X is the trajectory of a 'fluid particle', namely $\dot{X} = u(t, X)$.

When dissipation ($\nu \neq 0$) and stretching (the term $D_u\, \xi$) are taken into account, the whole problem is more difficult. Fluid particles keep moving following the velocity field, but in order to take into account dissipation, it is more convenient to single out the effect of diffusion, using the dynamics

$$dX = u(t, X)\,dt + \sqrt{2\nu}\,dB_t, \tag{10.7}$$

with B the three-dimensional standard Brownian motion. To take vortex stretching into account, we can imagine that when we compare vorticity at the two ends of a fluid particle trajectory, the cumulative effect of the deformation changes the vorticity size and direction. For instance, the vorticity is stretched when it is sufficiently aligned with the expanding directions of D_u. Since these directions change with time, $\xi(t, X(t))$ may undergo a complicated evolution with stretching, rotations, and contractions. Heuristic reasoning and numerical experiments show a predominance of the stretching mechanism,.

In Busnello et al. (2005) the three phenomena, transport, diffusion and stretching, are summarised by the following representation formula

$$\xi(t, x) = \mathbb{E}[V(t, 0)\xi(0, X_0)],$$

where $X(s)$ is the solution of (10.7) with *final* condition $X(t) = x$, and $V(r, s)$ is the solution of the 3×3 matrix equation

$$\begin{cases} \frac{d}{dr}V(r, s) = D_u(r, X(r))V(r, s), & r \in (s, t) \\ V(s, s) = I. \end{cases}$$

So far, the representation is incomplete, since V depends on the deformation tensor, hence from the velocity, that in turn can be reconstructed from the vorticity (with suitable decay). In Busnello et al. (2005) the reconstruction, through the Biot–Savart law, is also formulated as a probabilistic representation

$$u(x) = \int_0^\infty \frac{1}{2t} \boldsymbol{E}[\xi(x + W_t) \times W_t]\,dt,$$

where W is an additional three-dimensional standard Brownian motion. The representation we have described is implicit, since the formula for vorticity is given in terms of the velocity, and the formula for the velocity depends on the vorticity. Nevertheless, this allows us to formulate an alternative proof of a Beale–Kato–Majda like criterion given by Ponce (1985), and the local existence results for the vorticity given below.

Theorem 10.5.1. *Given $p \in [1, \frac{3}{2})$, $\alpha \in (0, 1)$, let $\xi_0 \in C_b^\alpha(\mathbb{R}^3, \mathbb{R}^3) \cap L^p(\mathbb{R}^3, \mathbb{R}^3)$. Then there are $\tau > 0$, depending only on $\|\xi_0\|_{C_b^\alpha \cap L^p}$, and a unique solution u of* (10.1) *given by the representation formulas above.*

Obviously, the smallness of τ can be replaced by the smallness of $\|\xi_0\|_{C_b^\alpha \cap L^p}$ to have global solutions. Also suitable external forces can be considered in the probabilistic formulation.

10.5.1.2 Eulerian–Lagrangian approach

An alternative representation formula that is based on the Eulerian–Lagrangian approach developed by Constantin (2001a,b) has been presented in Iyer (2006) and Constantin & Iyer (2008). The formula yields directly the velocity without passing through vorticity. On the one hand the effects of stretching are less apparent, and on the other there is a complete decoupling between diffusion and transport effects, namely the final formula for the viscous fluid is simply the expectation of the corresponding formula for inviscid flows.

The starting point is the *Weber formula* for an inviscid fluid (see Constantin, 2001a),

$$u = \Pi_L (\nabla A)^T (u_0(A)),$$

where Π_L is the Leray projector, $A_t = X_t^{-1}$ and for every $x \in \mathbb{R}^3$, and $X(\cdot; x)$ is the Lagrangian trajectory starting at x, that is, the solution of $\dot{X}_t = u(t, X_t)$ with initial condition x.

In order to take the effect of viscosity into account, one can consider the diffused Lagrangian trajectories (10.7), that is, $dX = u(t, X)\,dt + \sqrt{2\nu}dB$, with B three-dimensional standard Brownian motion. Define again the *back-to-label* map $A_t = X_t^{-1}$ (this time X is a stochastic flow), then the representation formula is given by

$$u = \mathbb{E}[\Pi_L (\nabla A)^T (u_0(A))].$$

One can immediately deduce an expectation for the vorticity,

$$\xi = \mathbb{E}[(\nabla X)\xi_0(A)],$$

and the formula can easily take an external force $f(t, x)$ into account,

$$u(t) = \mathbb{E}[\Pi_L (\nabla A_t)^T (u_0(A_t))] + \mathbb{E}\left[\Pi_L \int_0^t (\nabla X)^T f(s, X_s)\,ds\right].$$

Further development can be found in Constantin & Iyer (2006), Iyer & Mattingly (2008) and Iyer (2009).

10.5.1.3 Stochastic cascades via branching processes

Le Jan & Sznitman (1997), Blömker et al. (2007), Bakhtin (2006).

Unlike the previous representation formulas, the representation that will be examined in this section is completely explicit and the hidden randomness is provided by branching processes. The idea of using branching processes as the underlying engine of probabilistic representations is not new; see for instance Skorohod (1964), Ikeda, Nagasawa, & Watanabe (1968) and McKean (1975) where branching is coupled with diffusion, and the stochastic representation is derived directly in the physical space, so that the linear operator is limited to generators of diffusions and the nonlinearity is polynomial.

In Le Jan & Sznitman (1997) the authors are able to consider the Navier–Stokes nonlinearity by looking for a representation in Fourier space. Their method suggests a flow of the kinetic energy among scales governed by transition probabilities computed according to the (Fourier-transformed) nonlinearity, and hence called evocatively stochastic cascades. It turns out that their method is quite general and can handle a large class of semi-linear parabolic PDEs, or systems of PDEs. Here we follow the presentation in Blömker et al. (2007) and consider PDEs with periodic boundary conditions of the type

$$\partial_t u = Au + F(u) + f,$$

where A is an operator with a complete set of eigenfunctions, F is a polynomial nonlinearity in u and its derivatives, and f is a given driving function. The case of full space, as in Le Jan & Sznitman (1997), can be considered with similar ideas.

In short, the solution u is expanded into a Fourier series with respect to the eigen-functions of A. The PDE is transformed into a system of countably many ODEs for the Fourier coefficients. The solution of the system is represented by the expectation of a recursive functional over a tree of branching particles. The rules for branching, regeneration and death probabilities of particles arise from the particular PDE studied.

One major drawback of this stochastic representation is that it often fails to exist for large times t, although the solution to the PDE may still exist. The problem is that the recursive functional may fail to be integrable at some time. The work of Blömker et al. (2007) provides both a comparison equation whose finiteness implies integrability of the recursive functional, and a way to avoid non-integrability by suitably pruning the tree. A different approach, unfortunately working only for ODE, and based on resummation has been proposed by Morandin (2005).

Due to this problem it is not easy to tackle, by this approach, problems like the long time behaviour. This can be done only in special cases with small initial conditions and uniformly small forcing, as in Bakhtin (2006) for the Navier–Stokes equations with small bounded forcing.

We formulate the probabilistic representation for the two-dimensional Navier–Stokes in its vorticity formulation, in order to avoid the additional but harmless difficulty of vector-valued coefficients. One can proceed similarly for the three-dimensional case on the torus. In terms of Fourier series the equation reads

$$\dot{\xi}_k = -|k|^2 \xi_k + \sum_{m+n=k} \frac{k \cdot m^\perp}{|m|^2} \xi_m \xi_n + f_k,$$

where $(\xi_k)_{k \in \mathbb{Z}^2_*}$ are the Fourier coefficients of the vorticity with respect to the complex exponentials, $(f_k)_{k \in \mathbb{Z}^2_*}$ the coefficients of the forcing. We assume that $\chi_0 = f_0 = 0$. Set $\chi_k = |k|^\alpha \xi_k$ for some suitable $\alpha > \frac{1}{2}$, and define $\lambda_k = |k|^2$; for $c_{16} > 0$ we define

$$B_{kmn}(\chi, \chi') = \frac{k \cdot m^\perp}{|k \cdot m^\perp|} \chi \chi',$$

$$q_{kmn} = \frac{|k|^{\alpha-2}|k \cdot m^\perp|}{c_{16}|m|^{\alpha+2}|n|^\alpha}, \qquad \gamma_k = \frac{|k|^{\alpha-2}}{d_k} f_k,$$

for all k, m, $n \in \mathbb{Z}^2_*$ satisfying $k \cdot m^\perp \neq 0$ and $m + n = k$ (and zero otherwise). It is elementary to show that $q_k = \sum_{mn} q_{kmn} < \infty$ and that $q_k \to 0$ as $|k| \to \infty$. By choosing c_{16} large enough we have $q_k < 1$; we set $d_k = 1 - q_k$. The equations are recast in the following form,

$$\dot{\chi}_k = \lambda_k \Big(-\chi_k + c_{16} \sum_{m,n \in \mathbb{Z}^2_*} q_{kmn} B_{kmn}(\chi_m, \chi_n) + d_k \gamma_k \Big),$$

or, better, in its mild form

$$\chi_k(t) = e^{-\lambda_k t} \chi_k(0) + d_k \lambda_k \int_0^t e^{-\lambda_k(t-s)} \gamma_k(s) \, ds$$

$$+ c_{16} \lambda_k \int_0^t e^{-\lambda_k(t-s)} \sum_{m,n \in \mathbb{Z}^2_*} q_{kmn} B_{kmn}(\chi_m(s), \chi_n(s)) \, ds,$$

with $k \in \mathbb{Z}^2_*$. The constants $\lambda_k > 0$ are the rate of particle evolution), while q_{kmn} and d_k are the probabilities of branching and dying.

We describe first the branching tree. Define the labels set $\mathscr{I} = \bigcup_{n=0}^\infty \{1,2\}^n$, the history of a particle $\alpha = (\alpha_1, \dots, \alpha_n)$ can be read off by interpreting $\alpha_j = 1$ (or 2) as being child 1 (or 2) in a binary branching event at generation j. For fixed $k \in \mathbb{Z}^2_*$, a tree rooted at k is a system, indexed over $\mathscr{I}$, of particle positions $\mathbf{K}_\alpha$, birth τ_α^B and

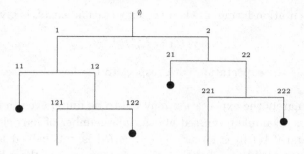

Fig. 10.6. A tree with branches and deaths ($\bullet$).

death τ_α^D times, defined inductively over the length of the labels. At the root $\mathbf{K}_\emptyset = k$, $\tau_\emptyset^B = 0$, and $\tau_\emptyset^D$ is exponential with rate λ_k. Given the tree with particles with n ancestors, each of these particles, with position say k', either die with probability $d_{k'}$, or disappear giving raise to two new independent particles, with positions m and n with probability $q_{k'mn}$. The new particles will have a lifespan distributed as independent exponential random variables with rates λ_m and λ_n.

Notice that by construction, given a branching particle giving raise to two particles at positions m and n, and conditional to its genealogy, the two sub-trees generated are independent and with the same distribution of trees rooted at m and n.

To ensure that the tree has only finitely many branches before a given time t, a sufficient condition is that $q_k \leq d_k$. We then define the evaluation operator along the tree. Fix a forcing functions γ, an initial condition $\chi(0)$ and a time $t > 0$. We define an evaluation map R_t recursively backwards along the tree. For a finite tree $\mathcal{T}$ produced with the above rules, start at time $s = t$ and work back to time $s = 0$: evaluate the initial condition $\chi(0)$ at any particles that are alive at time t, evaluate the forcing function $\gamma(s)$ at any particle that dies at time $s < t$, and apply the bi-linear operators at the times of branching events.

To understand how the evaluation works, let us consider the simple example of one possible position and the ODE $\dot{\chi} = -\chi + \frac{1}{2}\chi^2 + \frac{1}{2}f$, with f constant in time. We have $d = q = \frac{1}{2}$, $c_{16} = 1$ and the bi-linear operator is the usual product. Hence on a tree $\mathcal{T}$ the evaluation yields $R(\mathcal{T}) = \chi(0)^{A_t} f^{D_t}$, where A_t are the particles alive at time t and D_t are the particles dead by time t.

Consider again the general system we have discussed so far. For an initial condition $\chi(0) \in \ell^\infty(\mathbb{C}^r)$, and a forcing $\gamma \in L^\infty([0,T], \ell^\infty(\mathbb{C}^r))$, the representation formula, when the expectation exists, is given by

$$\chi_k(t) = \boldsymbol{E}_k[R_t],$$

where $\boldsymbol{E}_k$ is the expectation with respect to the law of a tree rooted at level k.

Unfortunately the expectation may fail to be finite, even in the seemingly simple example discussed above. The number of particles alive is Poisson, and if $|\chi(0)|$ is larger than 1, $\boldsymbol{E}[R_t]$ is not defined for t large enough (but possibly smaller than the existence time of the solution). A method based on pruning of trees was proposed by Blömker et al. (2007) to avoid this divergence problem.

10.5.2 Statistical topics

In this last section we briefly summarise some recent works associated by the common idea of applying statistical theories to estimate the values of parameters, or the distribution of the driving forces. These results are mainly justified, at least in their spirit, by applications to weather forecasting. This also justifies (together with better analytic estimates) the fact that the analysis is centred around the two-dimensional case.

In Cialenco & Glatt-Holtz (2011) the main aim is to give an estimate of the viscosity in the two-dimensional Navier–Stokes equations driven by noise, with periodic or Dirichlet boundary conditions, given the full observation of a path in a time interval $[0, T]$. The infinite-dimensional problem is not regular, namely the probability distributions, for different values of ν, are not mutually equivalent. The authors compute the maximum likelihood ratio estimator of spectral approximations of the problem. Galerkin approximations are needed to recover regularity. The estimators are weakly consistent and asymptotically normal, although tricky to be computed. The authors formulate two simplified estimators that are still weak consistent (although the rate of convergence is not clear) and depending only on a finite number of modes.

Hoang, Law, & Stuart (2013) discuss how to recover the driving force and the initial conditions, given noisy observations of the fluid. They based their analysis on the Bayesian approach, looking for the maximum a posteriori estimator (minimizing a least square problem), given that the prior distributions of initial condition and forcing are Gaussian, and the forcing is white in time.

Finally, the work Blömker et al. (2013) (see also Brett et al., 2013), deals with filtering to improve the accuracy of the estimate of the state of the system, in view of updating the posterior distribution.

References

Albeverio, S. & Belopolskaya, Y. (2003) Probabilistic approach to hydrodynamic equations, in *Probabilistic methods in fluids*. World Sci. Publ., River Edge, NJ, 1–21.

Albeverio, S. & Cruzeiro, A.B. (1990) Global flows with invariant (Gibbs) measures for Euler and Navier–Stokes two-dimensional fluids, *Comm. Math. Phys.* **129**, no. 3, 431–444.

Albeverio, S. & Ferrario, B. (2008) Some methods of infinite dimensional analysis in hydrodynamics: an introduction, in *SPDE in hydrodynamic: recent progress and prospects*. Lecture Notes in Math. **1942**, Springer, Berlin, 1–50.

Bakhtin, Y. (2006) Existence and uniqueness of stationary solutions for 3D Navier-Stokes system with small random forcing via stochastic cascades. *J. Stat. Phys.* **122**, no. 2, 351–360.

Bally V. & Caramellino, L. (2012) Regularity of probability laws by using an interpolation method. arXiv:1211.0052.

Barbato, D. & Morandin, F. (2013a) Positive and non-positive solutions for an inviscid dyadic model: well-posedness and regularity. *NoDEA Nonlinear Differential Equations Appl.* **20**, no. 3, 1105–1123.

Barbato, D. & Morandin, F. (2013a) *Stochastic inviscid shell models: well-posedness and anomalous dissipation. Nonlinearity* **26**, no. 7, 1919–1943.

Barbato, D., Flandoli F., & Morandin, F. (2010a) A theorem of uniqueness for an inviscid dyadic model. *C. R. Math. Acad. Sci. Paris* **348**, no. 9-10, 525–528.

Barbato, D., Flandoli F., & Morandin, F. (2010b) Uniqueness for a stochastic inviscid dyadic model. *Proc. Amer. Math. Soc.* **138**, no. 7, 2607–2617.

Barbato, D., Flandoli, F., & Morandin, F. (2011a) Anomalous dissipation in a stochastic inviscid dyadic model. *Ann. Appl. Probab.* **21** no. 6, 2424–2446.

Barbato, D., Flandoli, F., & Morandin, F. (2011b) Energy dissipation and self-similar solutions for an unforced inviscid dyadic model. *Trans. Amer. Math. Soc.* **363**, no. 4, 1925–1946.

Barbato, D., Morandin, F., & Romito, M. (2011c) Smooth solutions for the dyadic model. *Nonlinearity* **24**, no. 11, 3083–3097.

Barbato, D., Bianchi, L.A., Flandoli, F., & Morandin, F. (2013) A dyadic model on a tree. *J. Math. Phys.* **54**, no. 2, 021507, 20.

Barbato, D., Morandin, F., & Romito, M. (2014) Global regularity for a logarithmically supercritical hyperdissipative dyadic equation. *Dyn. Partial Differ. Equ.* **11**, no. 1, 39–52.

Bianchi, L.A. (2013) Uniqueness for an inviscid stochastic dyadic model on a tree. *Electron. Commun. Probab.* **18**, no. 8, 1–12.

Blömker, D. & Romito, M. (2009) Regularity and blow up in a surface growth model. *Dyn. Partial Differ. Equ.* **6**, no. 3, 227–252.

Blömker, D. & Romito, M. (2012) Local existence and uniqueness in the largest critical space for a surface growth model. *NoDEA Nonlinear Differential Equations Appl.* **19**, no. 3, 365–381.

Blömker, D. & Romito, M. (2013) Local existence and uniqueness for a two–

dimensional surface growth equation with space–time white noise. *Stoch. Anal. Appl.* **31**, no. 6, 1049–1076.

Blömker, D., Romito, M., & Tribe, R. (2007) A probabilistic representation for the solutions to some non-linear PDEs using pruned branching trees. *Ann. Inst. H. Poincaré Probab. Statist.* **43**, no. 2, 175–192.

Blömker, D., Flandoli, F., & Romito, M. (2009) Markovianity and ergodicity for a surface growth PDE. *Ann. Probab.* **37**, no. 1, 275–313.

Blömker, D., Law, K., Stuart, A.M. , & Zygalakis, K.C. (2013) Accuracy and stability of the continuous-time 3DVAR filter for the Navier-Stokes equation. *Nonlinearity* **26**, no. 8, 2193–2219.

Bogachev, V.I. (1998) *Gaussian measures.* Mathematical Surveys and Monographs **62**, American Mathematical Society, Providence, RI.

Bourgain, J. (1994) Periodic nonlinear Schrödinger equation and invariant measures. *Comm. Math. Phys.* **166**, no. 1, 1–26.

Bourgain, J. (1996) Invariant measures for the 2D-defocusing nonlinear Schrödinger equation. *Comm. Math. Phys.* **176**, no. 2, 421–445.

Brett, C.E.A. , Lam, K.F. , Law, K.J.H., McCormick,D.S., Scott, M.R., & Stuart, A.M. (2013) Accuracy and stability of filters for dissipative PDEs. *Phys. D* **245**, 34–45.

Brzeźniak, Z. Capiński, M., & Flandoli, F. (1991) Stochastic partial differential equations and turbulence. *Math. Models Methods Appl. Sci.* **1**, no. 1, 41–59.

Burdzy, K,. Mueller, C., & Perkins, E.A. (2010) Nonuniqueness for nonnegative solutions of parabolic stochastic partial differential equations. *Illinois J. Math.* **54**, no. 4, 1481–1507.

Busnello, B., Flandoli, F., & Romito, M. (2005) A probabilistic representation for the vorticity of a three-dimensional viscous fluid and for general systems of parabolic equations. *Proc. Edinb. Math. Soc. (2)* **48** no. 2, 295–336.

Burq, N. & Tzvetkov, N. (2008a) Random data Cauchy theory for supercritical wave equations. I. Local theory. *Invent. Math.* **173**, no. 3, 449–475.

Burq, N. & Tzvetkov, N. (2008b) Random data Cauchy theory for supercritical wave equations. II. A global existence result. *Invent. Math.* **173**, no. 3, 477–496.

Burq, N. & Tzvetkov, N. (2014) Probabilistic well-posedness for the cubic wave equation. *J. Eur. Math. Soc.,* **16**, no. 1, 1–30.

Busnello, B. (1999) A probabilistic approach to the two-dimensional Navier-Stokes equations. *Ann. Probab.* **27**, no. 4, 1750–1780.

Cheskidov, A. (2008) Blow-up in finite time for the dyadic model of the Navier-Stokes equations. *Trans. Amer. Math. Soc.* **360**, no. 10, 5101–5120.

Cheskidov, A., & Friedlander, S. (2009) The vanishing viscosity limit for a dyadic model. *Phys. D* **238**, no. 8, 783–787.

Cheskidov, A., Friedlander, S., & Pavlović, N. (2007) Inviscid dyadic model of turbulence: the fixed point and Onsager's conjecture. *J. Math. Phys.* **48**, no. 6, 065503, 16.

Cheskidov, A., Friedlander, S., & Pavlović, N. (2010) An inviscid dyadic model of turbulence: the global attractor. Discrete Contin. Dyn. Syst. **26** (2010), no. 3, 781–794.

Cialenco, I. & Glatt-Holtz, N. (2011) Parameter estimation for the stochastically perturbed Navier-Stokes equations. *Stochastic Process. Appl.* **121**, no. 4, 701–724.

Constantin, P. (2001a) An Eulerian-Lagrangian approach for incompressible fluids: local theory. *J. Amer. Math. Soc.* **14**, no. 2, 263–278.

Constantin, P. (2001b) An Eulerian-Lagrangian approach to the Navier-Stokes equations, *Comm. Math. Phys.* **216**, no. 3, 663–686.

Constantin, P. & Iyer, G. (2006) Stochastic Lagrangian transport and generalized relative entropies. *Commun. Math. Sci.* **4**, no. 4, 767–777.

Constantin, P. & Iyer, G. (2008) A stochastic Lagrangian representation of the three-dimensional incompressible Navier-Stokes equations. *Comm. Pure Appl. Math.* **61**, no. 3, 330–345.

Cruzeiro, A.B. & Shamarova, E. (2009) Navier-Stokes equations and forward-backward SDEs on the group of diffeomorphisms of a torus. *Stochastic Process. Appl.* **119**, no. 12, 4034–4060.

Da Prato, G. & Debussche, A. (2002) Two-dimensional Navier-Stokes equations driven by a space-time white noise. *J. Funct. Anal.* **196**, no. 1, 180–210.

Da Prato, G. & Debussche, A. (2003a) Ergodicity for the 3D stochastic Navier-Stokes equations. *J. Math. Pures Appl.* (9) **82**, no. 8, 877–947.

Da Prato, G. & Debussche, A. (2003b) Strong solutions to the stochastic quantization equations. *Ann. Probab.* **31**, no. 4, 1900–1916.

Da Prato, G. & Debussche, A. (2008) On the martingale problem associated to the 2D and 3D stochastic Navier-Stokes equations. *Atti Accad. Naz. Lincei Cl. Sci. Fis. Mat. Natur. Rend. Lincei (9) Mat. Appl.* **19**, no. 3, 247–264.

Da Prato, G. & Zabczyk, J. (1992) *Stochastic equations in infinite dimensions.* Encyclopedia of Mathematics and its Applications **44**, Cambridge University Press, Cambridge.

Da Prato, G. & Zabczyk, J. (1996) *Ergodicity for infinite-dimensional systems*, London Mathematical Society Lecture Note Series, **229**, Cambridge University Press, Cambridge.

Deng, C. & Cui, S. (2011a) Random-data Cauchy problem for the Navier-Stokes equations on $\mathbb{T}^3$. *J. Differential Equations* **251**, no. 4-5, 902–917.

Deng, C. & Cui, S. (2011b) Random-data Cauchy problem for the periodic Navier-Stokes equations with initial data in negative-order Sobolev spaces. arXiv:1103.6170.

Debussche, A. (2013) Ergodicity results for the stochastic Navier–Stokes equations: an introduction, in *Topics in mathematical fluid mechanics.* Lecture Notes in Math. **2073**, Springer, Heidelberg, Berlin, 2013, Lectures given at the C.I.M.E.–E.M.S. Summer School in applied mathematics held in Cetraro, September 6–11, 2010, Edited by F. Flandoli & H. Beirao da Veiga, 23–108.

De Marco, S. (2011) Smoothness and asymptotic estimates of densities for SDEs with locally smooth coefficients and applications to square root-type diffusions. *Ann. Appl. Probab.* **21**, no. 4, 1282–1321.

Debussche, A. & Odasso, C. (2006) Markov solutions for the 3D stochastic Navier-Stokes equations with state dependent noise. *J. Evol. Equ.* **6**, no. 2, 305–324.

Debussche, A. & Romito, M. (2014) Existence of densities for the 3D Navier–Stokes equations driven by Gaussian noise. *Probab. Theory Related Fields* **158**, no. 3-4, 575–596.

Engelbert, H. J. & Schmidt, W. (1985) On solutions of one-dimensional stochastic differential equations without drift. *Z. Wahrsch. Verw. Gebiete*

68, no. 3, 287–314.

Esposito, R., Marra, R., Pulvirenti, M., & Sciarretta, C. (1988) A stochastic Lagrangian picture for the three-dimensional Navier-Stokes equation. *Comm. Partial Differential Equations* **13**, no. 12, 1601–1610.

Esposito, R. & Pulvirenti, M. (1989) Three-dimensional stochastic vortex flows. *Math. Methods Appl. Sci.* **11**, no. 4, 431–445.

Flandoli, F. (2008) An introduction to 3D stochastic fluid dynamics, SPDE in hydrodynamic: recent progress and prospects, in *Lecture Notes in Math.* **1942**, Springer, Berlin. Lectures given at the C.I.M.E. Summer School held in Cetraro, August 29–September 3, 2005. Edited by Da Prato, G. & Röckner, M., 51–150.

Flandoli, F. (2011) *Random perturbation of PDEs and fluid dynamic models.* Lecture Notes in Mathematics **2015**, Springer, Heidelberg. Lectures from the 40th Probability Summer School held in Saint-Flour, 2010.

Flandoli, F. & Gątarek, D. (1995) Martingale and stationary solutions for stochastic Navier-Stokes equations. *Probab. Theory Related Fields* **102**, no. 3, 367–391.

Flandoli, F. & Maslowski, B. (1995) Ergodicity of the 2-D Navier-Stokes equation under random perturbations. *Comm. Math. Phys.* **172**, no. 1, 119–141.

Flandoli, F. & Romito, M. (2001) Statistically stationary solutions to the 3-D Navier-Stokes equation do not show singularities. *Electron. J. Probab.* **6**, no. 5, 1–15.

Flandoli, F. & Romito, M. (2002a)Partial regularity for the stochastic Navier-Stokes equations. *Trans. Amer. Math. Soc.* **354**, no. 6, 2207–2241.

Flandoli, F. & Romito, M. (2002b) *Probabilistic analysis of singularities for the 3D Navier-Stokes equations.* Proceedings of EQUADIFF, 10 (Prague, 2001), **127**, 211–218.

Flandoli, F. & Romito, M. (2006) Markov selections and their regularity for the three-dimensional stochastic Navier-Stokes equations. *C. R. Math. Acad. Sci. Paris* **343**, no. 1, 47–50.

Flandoli, F. & Romito, M. (2007) Regularity of transition semigroups associated to a 3D stochastic Navier-Stokes equation, in *Stochastic differential equations: theory and applications.* Interdiscip. Math. Sci., **2**, World Sci. Publ., Hackensack, NJ, (P.H. Baxendale & S.V. Lototski, eds.) 263–280.

Flandoli, F. & Romito, M. (2008) *Markov selections for the 3D stochastic Navier-Stokes equations.* Probab. Theory Related Fields **140**, no. 3-4, 407–458.

Flandoli, F., Gubinelli, M., Hairer, M., & Romito, M. (2008) Rigorous remarks about scaling laws in turbulent fluids. *Comm. Math. Phys.* **278**, no. 1, 1–29.

Flandoli, F. Gubinelli, M, & Priola, E. (2010) Well-posedness of the transport equation by stochastic perturbation. *Invent. Math.* **180** (2010), no. 1, 1–53.

Foiaş, C. Statistical study of Navier-Stokes equations. I, II. *Rend. Sem. Mat. Univ. Padova* **48**, 219–348; ibid. **49** , 9–123.

Foias, C., Manley, P., Rosa, R., & Temam, R. (2001) *Navier-Stokes equations and turbulence.* Encyclopedia of Mathematics and its Applications **83**, Cambridge University Press, Cambridge.

Fournier, N. & Printems, J. (2010) Absolute continuity for some one–dimensional processes. *Bernoulli* **16**, no. 2, 343–360.

Friedlander, S. & Pavlović, N. (2004) Blowup in a three-dimensional vector model for the Euler equations. *Comm. Pure Appl. Math.* **57**, no. 6, 705–725.

Fursikov, A.V. (1980) Some control problems and results related to the unique solvability of the mixed boundary value problem for the Navier-Stokes and Euler three-dimensional systems. *Dokl. Akad. Nauk SSSR* **252**, no. 5, 1066–1070.

Fursikov, A.V. (1981a) Control problems and theorems concerning unique solvability of a mixed boundary value problem for three-dimensional Navier-Stokes and Euler equations. *Mat. Sb. (N.S.)* **115**, no. 2, 281–306.

Fursikov, A.V. (1981b) On the question of the unique solvability of a three-dimensional Navier-Stokes system under almost all initial conditions. *Uspekhi Mat. Nauk* **36**, no. 2, 207–208.

Fursikov, A.V. (1983) Statistical extremal problems and unique solvability of the three-dimensional Navier-Stokes system under almost all initial conditions. *J. Appl. Math. Mech.* **46**, no. 5, 797–805.

Fursikov, A. V. (1984) Space-time moments and statistical solutions concentrated on smooth solutions of a three-dimensional Navier-Stokes system or a quasilinear parabolic system. *Dokl. Akad. Nauk SSSR* **274**, no. 3, 548–553.

Fursikov, A.V. (1987) On the uniqueness of the solution of a chain of moment equations that correspond to a three-dimensional Navier–Stokes system. *Mat. Sb. (N.S.)* **134(176)**, no. 4, 472–495.

Fujitam H. & Kato, T. (1964) On the Navier–Stokes initial value problem. I. *Arch. Rational Mech. Anal.* **16**, 269–315.

Gubinelli, M., Imkeller, P., & Perkowski, N. (2013) Paracontrolled distributions and singular PDEs. arXiv:1210.2684.

Guermond, J.-L. (2006) Finite-element-based Faedo-Galerkin weak solutions to the Navier-Stokes equations in the three-dimensional torus are suitable. *J. Math. Pures Appl. (9)* **85**, no. 3, 451–464.

Hairer, M. (2013) Solving the KPZ equation. *Ann. of Math. (2)* **178**, no. 2, 559–664.

Hairer, M. (2014b) *Singular stochastic PDEs*, in Proceedings of the ICM.

Hairer, M. (2015a) Introduction to Regularity Structures. *Braz. Jour. Prob. Stat.* **29**, 175–210.

Hairer, M. (2015b) A theory of regularity structures. *Invent. Math.* **198**, 269–504.

Hairer, M. & Mattingly, J.C. (2006) Ergodicity of the 2D Navier-Stokes equations with degenerate stochastic forcing. *Ann. of Math. (2)* **164**, no. 3, 993–1032.

Hayashi, M., Kohatsu-Higa, A., & Yûki, G. (2013a) Local Hölder continuity property of the densities of SDEs with singular drift coefficients. *J. Theoret. Probab.* **26**, 1117–1134.

Hayashi, M., Kohatsu-Higa, A., & Yûki, G. (2013b) Local Hölder continuity property of the densities of solutions of SDEs with singular coefficients. *J. Theoret. Probab.* **26**, no. 4, 1117–1134.

Hoang, V.H., Law, K.J.H, & Stuart, A.M. Determining white noise forcing from eulerian observations in the Navier–Stokes equation. *Stoch. Partial Differ. Equ. Anal. Comput.* **2**, 233–261.

Ikeda, N., Nagasawa, M., & Watanabe, S. (1968) Branching Markov processes. I. *J. Math. Kyoto Univ.* **8**, 233–278.

Iyer, G. (2006) A stochastic perturbation of inviscid flows. *Comm. Math. Phys.* **266**, no. 3, 631–645.

Iyer, G (2009) A stochastic Lagrangian proof of global existence of the Navier-Stokes equations for flows with small Reynolds number. *Ann. Inst. H. Poincaré Anal. Non Linéaire* **26**, no. 1, 181–189.

Iyer, G. & Mattingly, J. (2008) A stochastic-Lagrangian particle system for the Navier-Stokes equations. *Nonlinearity* **21**, no. 11, 2537–2553.

Kato, T. (1984) Strong L^p-solutions of the Navier–Stokes equation in $\mathbf{R}^m$, with applications to weak solutions. *Math. Z.* **187**, no. 4, 471–480.

Katz, N.H. & Pavlović, N. (2005) Finite time blow-up for a dyadic model of the Euler equations. *Trans. Amer. Math. Soc.* **357**, no. 2, 695–708.

Kohatsu-Higa, A. & Tanaka, A. (2012) A Malliavin calculus method to study densities of additive functionals of SDE's with irregular drifts. *Ann. Inst. Henri Poincaré Probab. Stat.* **48**, no. 3, 871–883.

Krylov, N.V. (1973) The selection of a Markov process from a Markov system of processes, and the construction of quasidiffusion processes. *Izv. Akad. Nauk SSSR Ser. Mat.* **37**, 691–708.

Krylov, N.V. (2004) On weak uniqueness for some diffusions with discontinuous coefficients. *Stochastic Process. Appl.* **113**, no. 1, 37–64.

Krylov, N.V. & Röckner, M. (2005) Strong solutions of stochastic equations with singular time dependent drift. *Probab. Theory Related Fields* **131**, no. 2, 154–196.

Le Jan, Y. & Sznitman, A.S. (1997) Stochastic cascades and 3-dimensional Navier-Stokes equations. *Probab. Theory Related Fields* **109**, no. 3, 343–366.

Lemarié-Rieusset, P.G. (2002) *Recent developments in the Navier-Stokes problem*, Chapman & Hall/CRC Research Notes in Mathematics **431**, Chapman & Hall/CRC, Boca Raton, FL.

Lions, P.L. (1996)*Mathematical topics in fluid mechanics. Vol. 1.* Oxford Lecture Series in Mathematics and its Applications **3**, The Clarendon Press Oxford University Press, New York.

McKean, H. P. (1975) Application of Brownian motion to the equation of Kolmogorov-Petrovskii-Piskunov. *Comm. Pure Appl. Math.* **28**, no. 3, 323–331.

Morandin, F. (2005) A resummed branching process representation for a class of nonlinear ODEs. *Electron. Comm. Probab.* **10**, 1–6.

Mattingly, J.C. & Pardoux, E. (2006) Malliavin calculus for the stochastic 2D Navier-Stokes equation. *Comm. Pure Appl. Math.* **59**, no. 12, 1742–1790.

Mikulevicius, R. & Rozovskii, B.L. (2004) Stochastic Navier-Stokes equations for turbulent flows. *SIAM J. Math. Anal.* **35**, no. 5, 1250–1310.

Mikulevicius, R. & Rozovskii, B.L. (2005) Global L_2-solutions of stochastic Navier–Stokes equations. *Ann. Probab.* **33**, no. 1, 137–176.

Mueller, C., Mytnik, L., & Edwin, P. (2012) Nonuniqueness for a parabolic SPDE with $\frac{3}{4} - \epsilon$ hölder diffusion coefficients. *Ann. Probab.* **42**, 2032–2112.

Nahmod, A.R., Pavlović, & Staffilani, G. (2013) Almost sure existence of global weak solutions for supercritical Navier-Stokes equations. *SIAM J. Math. Anal.* **45**, no. 6, 3431–3452.

Nualart, D. (2006) *The Malliavin calculus and related topics.* Probability and its Applications (New York), Springer-Verlag, Berlin, Berlin.

Odasso, C. (2007) Exponential mixing for the 3D stochastic Navier-Stokes equations. *Comm. Math. Phys.* **270**, no. 1, 109–139.

Ponce, G. (1985) Remarks on a paper: "Remarks on the breakdown of smooth solutions for the 3-D Euler equations" by J. T. Beale, T. Kato & A. Majda. *Comm. Math. Phys.* **98**, no. 3, 349–353.

Romito, M. (2004) Ergodicity of the finite dimensional approximation of the 3D Navier-Stokes equations forced by a degenerate noise. *J. Statist. Phys.* **114**, no. 1-2, 155–177.

Romito, M. (2005) A geometric cascade for the spectral approximation of the Navier-Stokes equations, in *Probability and partial differential equations in modern applied mathematics*. IMA Vol. Math. Appl. **140**, Springer, New York, 197–212.

Romito, M. (2006) Some examples of singular fluid flows. *oDEA Nonlinear Differential Equations Appl.* **13**, no. 1, 67–89.

Romito, M. (2008) Analysis of equilibrium states of Markov solutions to the 3D Navier-Stokes equations driven by additive noise. *J. Stat. Phys.* **131**, no. 3, 415–444.

Romito, M. (2010a) An almost sure energy inequality for Markov solutions to the 3D Navier-Stokes equations, in *Stochastic Partial Differential Equations and Applications*. Quad. Mat. **25**, Dept. Math., Seconda Univ. Napoli, Caserta, 243–255.

Romito, M. (2010b) Existence of martingale and stationary suitable weak solutions for a stochastic Navier–Stokes system. *Stochastics* **82**, no. 1-3, 327–337.

Romito, M. (2011a) Critical strong Feller regularity for Markov solutions to the Navier–Stokes equations. *J. Math. Anal. Appl.* **384**, no. 1, 115–129.

Romito, M. (2011b) The Martingale problem for Markov solutions to the Navier-Stokes equations, in Seminar on Stochastic Analysis, Random Fields and Applications VI. *Progr. Probab.* **63**, Birkhäuser/Springer Basel AG, Basel, Basel, 227–244.

Romito, M. (2013) *Densities for the Navier–Stokes equations with noise*. Lecture notes for the "Winter school on stochastic analysis and control of fluid flow", School of Mathematics of the Indian Institute of Science Education and Research, Thiruvananthapuram (India).

Romito, M. (2014a) Densities for the 3D Navier–Stokes equations driven by degenerate Gaussian noise, in preparation.

Romito, M. (2014b) Density of the energy for the Navier–Stokes equations with noise, in preparation.

Romito, M. (2014c) Unconditional existence of densities for the Navier-Stokes equations with noise, in *RIMS Kôkyûroku* **1905**, 5–17.

Romito, M. (2014e) Uniqueness and blow-up for a stochastic viscous dyadic model, *Probab. Theory Related Fields* **158**, no. 3-4, 895–924.

Romito, M. (2015) Time regularity of the densities for the Navier–Stokes equations with noise. *J. Evol. Eq*, to appear.

Romito, M. & Xu, L. (2011) *Ergodicity of the 3D stochastic Navier-Stokes equations driven by mildly degenerate noise*. Stochastic Process. Appl. **121**, no. 4, 673–700.

Scheffer, V. (1977) Hausdorff measure and the Navier-Stokes equations. *Comm. Math. Phys.* **55**, no. 2, 97–112.

Scheutzow, M. (1993) Stabilization and destabilization by noise in the plane. *Stochastic Anal. Appl.* **11**, no. 1, 97–113.

Skorohod, A.V. (1964) Branching diffusion processes. *Teor. Verojatnost. i Primenen* **9**, 492–497.

Stroock, D.W. & Varadhan, S.R.S. (1979) Multidimensional diffusion processes. Grundlehren der Mathematischen Wissenschaften Fundamental Principles of Mathematical Sciences, **233**, Springer-Verlag, Berlin, New York.

Tao, T. (2016) Finite time blowup for an averaged three-dimensional Navier-Stokes equation. *J. Amer. Math. Soc.*, to appear.

Temam, R. (1995) *Navier-Stokes equations and nonlinear functional analysis*, CBMS-NSF Regional Conference Series in Applied Mathematics, **66**, Society for Industrial and Applied Mathematics (SIAM), Philadelphia, PA.

Triebel, H. (1983) *Theory of function spaces.* Monographs in Mathematics, **78**, Birkhäuser Verlag, Basel.

Triebel, H. (1992) *Theory of function spaces II.* Monographs in Mathematics, **84**, Birkhäuser Verlag, Basel.

Višik, M.I. & Fursikov, A.V. (1988) *Mathematical problems of statistical hydromechanics.* Mathematics and its applications (Soviet Series), Kluwer Academic Publishers, Dordrecht, Boston, London.

Zhang, T. & Fang, D. (2011) Random data Cauchy theory for the incompressible three dimensional Navier-Stokes equations. *Proc. Amer. Math. Soc.* **139**, no. 8, 2827–2837.

Printed in the United States
By Bookmasters